Contemporary Trends and Issues in Science Education

Volume 69

The book series Contemporary Trends and Issues in Science Education provides a forum for innovative trends and issues impacting science education. Scholarship that focuses on advancing new visions, understanding, and is at the forefront of the field is found in this series. Authoritative works based on empirical research and/or conceptual theory from disciplines including historical, philosophical, psychological and sociological traditions are represented here. Our goal is to advance the field of science education by testing and pushing the prevailing sociocultural norms about teaching, learning, research and policy. Book proposals for this series may be submitted to the Publishing Editor: Claudia Acuna E-mail: Claudia.Acuna@springer.com

Paul Nnanyereugo Iwuanyanwu

Empowering STEM Thinkers Through Argumentation

A Framework for Critical Practice

Springer

Paul Nnanyereugo Iwuanyanwu
Northwest University
Potchefstroom, South Africa

ISSN 1878-0482 ISSN 1878-0784 (electronic)
Contemporary Trends and Issues in Science Education
ISBN 978-3-032-27905-7 ISBN 978-3-032-27906-4 (eBook)
https://doi.org/10.1007/978-3-032-27906-4

This Springer imprint is published by the registered company Springer Nature Switzerland AG
The registered company address is: Gewerbestrasse 11, 6330 Cham, Switzerland

I dedicate this work to my beloved wife, Mimica Iwuanyanwu, and our children, Paul Jr., Olisa, and Authur Chukwuemeka Iwuanyanwu. Your unwavering love, patience, and encouragement have been the foundation upon which this project was built. I also honour my parents, Dr. Festus Iwuanyanwu and Mrs. Magdaline Iwuanyanwu, whose guidance, values, and steadfast support have shaped the person and scholar I have become. This book is a testament to the enduring power of family as both anchor and inspiration in the pursuit of knowledge.

Preface

We live in a moment where the pace of technological change challenges not only what we know but how we know. STEM professionals and students confront data that is vast, models that conflict, and tools that amplify human capability while testing ethical boundaries. In such a world, technical proficiency alone is insufficient. What is required is the disciplined capacity to reason, to weigh evidence, to interrogate assumptions, and to justify decisions amid uncertainty. This handbook is written to guide that reasoning. It invites us to cultivate argumentation as the cognitive and ethical architecture of STEM, the practice through which knowledge becomes reliable, decisions become responsible, and innovation becomes accountable. STEM work is iterative, contested, and interpretive. We encounter ambiguity, competing models, and unpredictable consequences across scientific, technological, and engineering contexts. Whether evaluating AI outputs, interpreting CRISPR-mediated genetic variations, navigating IoT vulnerabilities, or designing responsive robotic systems, we are engaged in structured reasoning. Argumentation is central to this work. Through it, we make thinking visible, surfacing assumptions, examining evidence, and defending claims within communities of practice. It is the tool that transforms knowledge into understanding and action.

Argumentation as presented here is more than debate or persuasion. It is a disciplined structure linking analysis, dialogue, and communication. Analytical argumentation traces connections between premises and conclusions. Dialectical argumentation exposes ideas to counterpositions that refine them. Rhetorical argumentation crafts explanations that are coherent, compelling, and contextually grounded. Together, these modes form the cognitive toolkit every STEM practitioner must cultivate to navigate complexity, innovation, and ethical responsibility. This handbook integrates philosophical reflection, practical frameworks, real-world examples, and design-oriented applications to show how argumentation functions as both method and mindset. It is offered not as a definitive guide but as an invitation to ask better questions, think critically, communicate clearly, and act wisely. As we engage with intelligent machines and complex knowledge systems, reasoned argument remains our bridge between knowledge and understanding, data and judgement, and technology and responsibility. Readers are invited to move beyond observation to active inquiry, reasoning alongside these pages, and carrying forward the practices that sustain thoughtful, reflective, and ethically grounded STEM work.

Potchefstroom, South Africa Paul Nnanyereugo Iwuanyanwu

Acknowledgments

I am deeply grateful to my mentors, Professor Emeritus Meshach Ogunniyi and Prof Neal Petersen, whose wisdom, guidance, and mentorship have profoundly influenced my professional development. I extend sincere appreciation to the Digital Art and Design team at North-West University for translating my ideas into visual illustrations that bring this work to life. My heartfelt thanks also go to my students, Buhle Kgatla and Ratheko Clifford, whose creative artistic skills transformed my conceptual ideals into compelling visual illustrations. Each of you has contributed to the realisation of this book, reminding me that scholarship is both a personal endeavour and a collaborative journey.

Competing Interests The author has no competing interests to declare that are relevant to the content of this manuscript.

About the Book

This book offers a comprehensive framework for developing critical, creative, and ethical minds in STEM education and professional practice. It addresses the urgent need for a new generation of STEM thinkers capable of navigating rapidly evolving technologies, complex societal challenges, and ethically demanding decision-making environments. The text argues that STEM competence extends beyond mastery of content to include disciplined reasoning, effective communication, collaborative engagement, digital literacy, and responsible knowledge practices.

Across nine interconnected chapters, the book integrates conceptual foundations with practical applications. It begins by establishing argumentation as the core cognitive and ethical practice that supports knowledge construction, critical judgment, and reflective reasoning. The chapters then examine how argumentation operates in STEM contexts, detailing components such as claims, evidence, warrants, counter-arguments, and reasoning strategies. Subsequent chapters explore instructional and professional applications, including designing ethically responsible learning experiences, applying scientific practices for evidence-based reasoning, and integrating engineering design thinking as a creative, iterative, and morally informed process. The final chapter situates learning within networked digital environments, showing how connectivist approaches, underpinned by structured argumentation, foster collaborative knowledge construction, innovation, and ethical participation in complex STEM ecosystems.

The book presents STEM learning as a holistic capability rather than the accumulation of facts or procedures. It emphasises the cultivation of sophisticated habits of mind that enable students, educators, researchers, and practitioners to interpret evidence critically, design effective solutions, evaluate information rigorously, collaborate productively, and act responsibly in technological contexts. By providing conceptual clarity, pedagogical principles, practical strategies, and research-informed insights, the book advances a human-centred vision for STEM education. It equips readers to prepare for a future shaped by complexity, interconnectivity, and continual transformation.

Contents

About the Author

Paul Nnanyereugo Iwuanyanwu (PhD) is a senior lecturer in the School of Mathematics, Science and Technology Education at North-West University. His research centres on STEM education, argumentation, indigenous knowledge systems, culture studies, ethical reasoning, and the design of pedagogical frameworks that support critical, creative, and responsible engagement with emerging technologies. He has published widely in top journals and book volumes and has received multiple academic excellence awards recognising his contributions to educational practice and didactics. Dr. Iwuanyanwu's work integrates conceptual, empirical, and practice-based insights to strengthen STEM teaching and learning across diverse educational contexts and to promote human-centred engagement in an increasingly complex technological world.

List of Abbreviations

4IR	Fourth Industrial Revolution
AI	Artificial Intelligence
AR	Augmented Reality
CAT	Contiguity Argumentation Theory
CRISPR	Clustered Regularly Interspaced Short Palindromic Repeats
IoT	Internet of Things
NAE	National Academy of Engineering
NRC	National Research Council
OECD	Organisation for Economic Co-operation and Development
STEM	Science, Technology, Engineering and Mathematics
TAP	Toulmin's Argument Pattern
UNECOSOC	United Nations Economic and Social Council
UNDESA	United Nations Department of Economic and Social Affairs
VR	Virtual Reality
WEF	World Economic Forum

List of Figures

List of Tables

Chapter 1
Introduction: Why Argumentation Matters in STEM

Before we measure, model, or manipulate the world, we must confront a more unsettling question: *What grants our claims the authority to be believed?* For if STEM is often portrayed as the domain of precision and certainty, its true power lies not in the accumulation of facts, but in the disciplined struggle to justify them. Every equation we trust, every model we deploy, and every technological system we design stands upon an invisible architecture of reasoning, an architecture forged through argument. We are therefore not merely discoverers of knowledge; we are its architects and its critics, bound by a responsibility to ensure that what we claim to know can withstand scrutiny, doubt, and revision. In this sense, argumentation is not an accessory to STEM practice, it is its moral and epistemic foundation.

We begin by acknowledging that the syllogistic method articulated by Aristotle in *Prior Analytics* remains the primordial framework that gave form, legitimacy, and discipline to deductive reasoning itself. His argument structure and enduring epistemic authority will forever retain their significance as the foundation on which other argumentation frameworks will evolve. Knowledge in any discipline is not simply discovered; it is earned through argument. Argument transforms belief from subjective conviction into epistemically legitimate understanding, creating the conditions under which claims can be publicly justified, critiqued, and refined. Across science, technology, engineering, and mathematics (STEM), professional practice is increasingly defined not only by what practitioners know, but by how they reason about, defend, and revise that knowledge in accountable and transparent ways. This is especially evident as frontier technologies, such as artificial intelligence, CRISPR gene editing, autonomous robotics, and IoT systems, introduce new layers of uncertainty, complexity, and ethical consequence.

It is precisely under these conditions of uncertainty, complexity, and ethical consequence that we come to recognise argumentation not as an abstract philosophical exercise, but as the living practice through which STEM knowledge is constructed, justified, and made accountable. Engineers argue for design trade-offs under

P. N. Iwuanyanwu, *Empowering STEM Thinkers Through Argumentation*, Contemporary Trends and Issues in Science Education 69,
https://doi.org/10.1007/978-3-032-27906-4_1

material, economic, safety, and ethical constraints, often involving algorithmic decision-making, autonomous systems, or human–machine interaction. Scientists defend claims through evidence, peer review, and methodological transparency, engaging in debates over the validity, reproducibility, and societal implications of gene-editing technologies and data-intensive research. Technologists negotiate competing models, risks, and values embedded in increasingly complex socio-technical systems, including AI-driven platforms and interconnected infrastructures. Mathematicians justify generalisations through proof, counterexample, and formal reasoning, providing the theoretical foundations for emerging computational and algorithmic methods. In each of these domains, argumentation is not ancillary to practice; it is constitutive of it (Gainsburg et al., 2016; Osborne, 2010).

We therefore proceed from the shared premise that empowering STEM thinkers entails inducting them into the epistemic and ethical practices of reasoned argument, especially as professionals must justify decisions that carry profound technical, social, and moral implications. In this sense, argumentation is the mechanism through which knowledge claims are warranted, contested, refined, and ultimately legitimised within disciplinary and professional communities confronting rapidly evolving technological frontiers. For instance, when students, practitioners and emerging professionals engage in argumentation, they move from reproducing established knowledge to participating in knowledge-building enterprises that shape how emerging technologies are designed, governed, and responsibly deployed (Jiménez-Aleixandre & Erduran, 2007; Kelly et al., 2007).

Figure 1.1 visually illustrates argumentation as the central connective tissue of STEM practice. At the core of the diagram, argumentation is shown as the organising epistemic activity through which STEM practice flows. The multiple directional arrows radiating outward signify that argumentation is not a linear or isolated act,

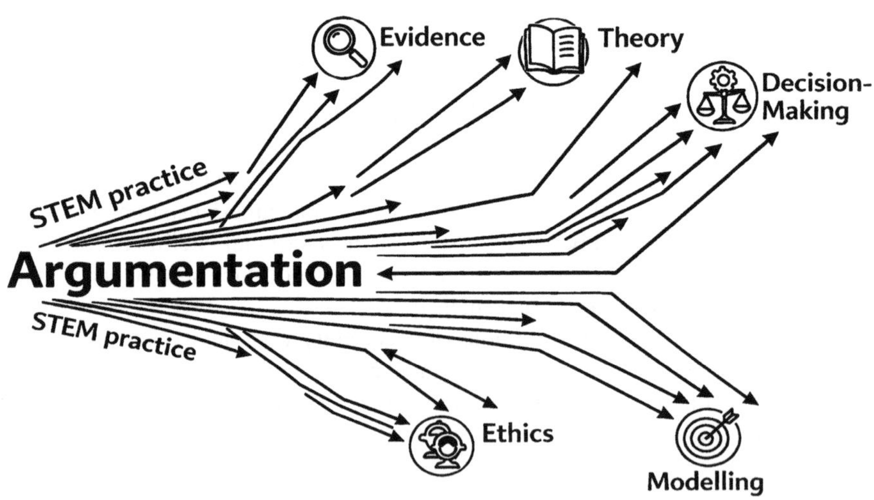

Fig. 1.1 Argumentation as the connective tissue of STEM practice

but a dynamic, iterative process that simultaneously draws from and feeds into key dimensions of professional STEM work. The upper trajectories illustrate how argumentation connects evidence and theory, reflecting how scientists and mathematicians justify claims through empirical data, peer review, formal proof, and theoretical coherence, including in data-rich and computationally intensive contexts. These pathways emphasise that evidence and theory do not stand alone; their legitimacy is established through reasoned justification, critique, and rebuttal. The rightward extension toward decision-making captures the professional reality that STEM arguments ultimately inform judgments under conditions of uncertainty, whether in engineering design trade-offs, assessments of autonomous system behaviour, governance of gene-editing technologies, or policy-relevant scientific and technological advice.

1.1 Philosophical Foundations of Argumentation as Epistemic Practice

This book adopts a philosophical stance that treats argumentation as an epistemic practice rather than a rhetorical skill. From this perspective, argumentation is a socially situated process through which knowledge is constructed, evaluated, and made credible. Toulmin's model of argument (claims, data, warrants, backings, qualifiers, and rebuttals) provides a foundational analytic lens for understanding how reasoning operates in real-world contexts where certainty is rare and judgement is required (Toulmin, 2003). In STEM disciplines, the elements of argument take distinctive forms. Evidence may consist of empirical data, simulations, mathematical derivations, or design artefacts.

Warrants may be theoretical principles, mathematical theorems, engineering heuristics, or modelling assumptions. Rebuttals often arise from anomalous data, boundary conditions, ethical concerns, or competing stakeholder priorities. What unifies these diverse instantiations is a shared commitment to reason-giving, critique, and accountability to disciplinary norms (Duschl & Osborne, 2002; Sampson & Clark, 2008). From a practice-theoretical perspective, argumentation is learned through participation. Novices become competent STEM practitioners not by memorising rules of argument, but by engaging in authentic practices where claims must withstand scrutiny. This view aligns with sociocultural theories of learning, which emphasise discourse, tools, and community norms as central to epistemic development (Gee, 2015; Nussbaum, 2011; Ogunniyi, 2007).

Figure 1.2 illustrates Toulmin's argument pattern to show how claims, evidence, warrants, and rebuttals take discipline-specific forms in science, mathematics, engineering, and technology, thereby reinforcing the idea that argumentation is not a generic rhetorical template but a situated mode of knowing. In the science column, the argument centres on an explanatory claim about biological behaviour. Empirical observations of beewolves' hunting practices function as evidence, while biological

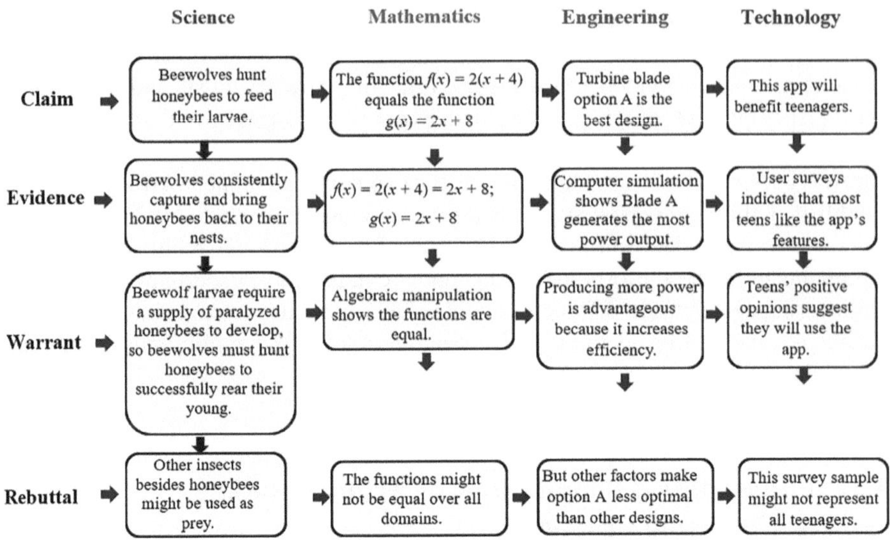

Fig. 1.2 Framing arguments across STEM fields

knowledge about larval development provides the warrant. The rebuttal acknowledges alternative explanations, illustrating how scientific knowledge remains provisional and open to critique. This aligns with the emphasis on argumentation as a mechanism for evaluating evidence and managing uncertainty in empirical inquiry.

The mathematics column exemplifies argumentation in a formal epistemic register. Here, the claim of functional equivalence is supported by symbolic manipulation as evidence, and algebraic rules serve as warrants. The rebuttal highlights domain constraints, demonstrating that even in mathematics, often perceived as certain, arguments remain conditional on defined assumptions and scopes. This reflects Toulmin's insight that warrants and rebuttals are essential even in highly formalised disciplines (Toulmin, 2003). In engineering, the diagram foregrounds argumentation as decision-oriented reasoning under constraints. The claim that a particular turbine blade design is optimal is supported by simulation data, with efficiency principles operating as warrants. The rebuttal introduces competing criteria and contextual factors, capturing the inherently evaluative and comparative nature of engineering judgement. This illustrates how engineering arguments integrate technical performance with broader considerations of feasibility and optimisation.

The technology column extends argumentation into socio-technical contexts. Claims about user benefit are grounded in survey data, with anticipated user behaviour functioning as a warrant. The rebuttal draws attention to sampling limitations, underscoring how technological decisions depend on negotiated interpretations of data, values, and stakeholder perspectives. This reinforces the chapter's argument that epistemic accountability in STEM includes responsiveness to social and ethical dimensions. Taken together, the figure demonstrates that while the structure of argumentation remains recognisable across disciplines, its epistemic substance varies according to disciplinary norms, tools, and purposes. By making these variations visible, the diagram supports the practice-theoretical view that students and professional novices develop STEM expertise by participating in authentic argumentative practices where claims must be justified, challenged, and refined within communities of practice (e.g., Duschl & Osborne, 2002; Ogunniyi, 2007; Sampson & Clark, 2008).

1.2 Learning Argumentation and Arguing to Learn

Research across STEM education demonstrates a crucial shift in emphasis that regards argumentation not only an outcome of learning, but a driver of learning. When learners engage in argumentation, they externalise reasoning, confront uncertainty, and test ideas against evidence and alternative explanations. This process supports conceptual change, epistemic awareness, and metacognitive growth (Asterhan & Schwarz, 2007; Erduran et al., 2004; Ryu & Sandoval, 2012). We therefore distinguish between learning argumentation, developing the skills, structures, and norms of constructing and evaluating arguments, and arguing to learn (Osborne, 2010), using argumentation as a means for understanding complex STEM concepts (Iwuanyanwu, 2020). Both are essential. Learning argumentation equips individuals with transferable reasoning tools, while arguing to learn deepens disciplinary understanding and professional judgement (Berland & McNeill, 2010; Osborne et al., 2004).

Let us now turn to Fig. 1.3 and Table 1.1, which work together to make visible the conceptual distinction between learning argumentation and arguing to learn. Read together, they help us see how argumentation functions both as an object of instruction and as a powerful engine for learning across core STEM practices, scientific inquiry, mathematical proof, and engineering design. Figure 1.3 invites us to consider two complementary orientations to argumentation. On the left, learning argumentation is foregrounded through disciplinary artefacts, laboratory equipment, symbolic mathematical representations, and design schematics. These visual cues draw our attention to situations in which argumentation itself becomes the focus of learning. Here, learners are apprenticed into the norms, structures, and evaluative criteria of disciplinary reasoning: (1) how claims are constructed, (2) how evidence is weighed, and (3) how justifications are judged for coherence and adequacy. Argumentation, in this orientation, is made explicit, named, and deliberately taught as a transferable epistemic practice.

On the right, arguing to learn shifts our gaze from structure to use. The emphasis moves to dialogue, collaboration, and joint problem-solving, where individuals employ argumentation as a means for making sense of disciplinary ideas. In these contexts, argumentation mediates learning by externalising reasoning, surfacing uncertainty, and enabling individuals to test and refine ideas in interaction with others. What we see here aligns with the research base discussed earlier: when individuals argue, they do not merely display what they know; they actively

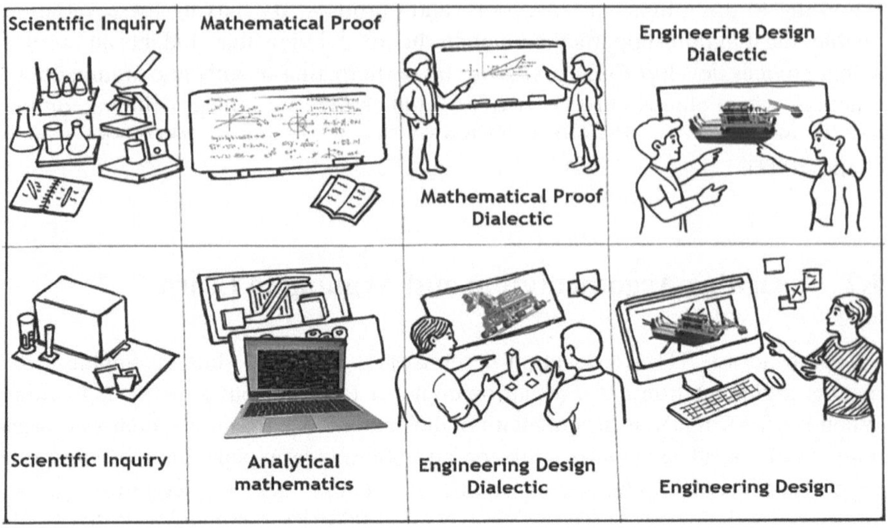

Fig. 1.3 Learning argumentation and arguing to learn across STEM practices

Table 1.1 Distinguishing learning argumentation from arguing to learn in STEM contexts

Context	Learning Argumentation (learning *about* argumentation)	Arguing to Learn (learning *through* argumentation)
Scientific inquiry	Practitioners and emerging scientists analyze and critique competing scientific explanations (e.g., evaluating claims, evidence, and warrants explaining why plants grow toward light) to understand the structure and quality of scientific arguments.	Practitioners and learners-in-training engage in discussion and debate about why plants grow toward light, using evidence from experiments to construct explanations and deepen conceptual understanding of phototropism.
Mathematical proof	Mathematicians and those developing proof expertise examine whether the steps in a proof (e.g., of the Pythagorean theorem) constitute valid justifications, focusing on logical structure, definitions, and accepted rules of inference.	Participants argue for the correctness of each step in a proof, justifying transitions and responding to counterarguments to strengthen their understanding of the underlying mathematical concepts.
Engineering design	Engineers and engineering trainees critique alternative design proposals (e.g., different bridge or turbine designs) by evaluating how well each argument is supported by performance data, constraints, and design criteria.	Engineers argue for and against design choices while iterating on prototypes, using evidence from testing and trade-off analysis to develop deeper understanding of engineering principles and constraints.

construct new understanding through critique, justification, and response (Driver et al., 2000; Sampson & Clark, 2008).

Table 1.1 grounds this distinction in concrete disciplinary examples. In scientific inquiry, learning argumentation involves critiquing explanations and evaluating the strength of evidence, whereas arguing to learn engages individuals in substantive discussions about phenomena, such as why plants lean toward light, in order to deepen conceptual understanding. In mathematical proof, the focus shifts from judging the validity of justificatory steps to using justification itself as a means of internalising mathematical principles. In engineering design, individuals move from critiquing alternative solutions against explicit criteria to using comparative argumentation to develop deeper insight into structural and functional relationships.

Taken together, Fig. 1.3 and Table 1.1 sharpen the distinction between learning argumentation and arguing to learn in STEM contexts, highlighting their complementary roles in developing robust STEM reasoning. Learning argumentation provides practitioners and emerging professionals with the normative tools of disciplinary reasoning, how claims are warranted, how evidence is evaluated, and how judgements are justified within professional communities. Arguing to learn, in turn, situates these tools within authentic STEM practice, where understanding is developed through inquiry, critique, and decision-making rather than through the passive accumulation of information. As we move through the chapters that follow, this dual emphasis remains foundational. It informs how we conceptualise curriculum design, frame professional learning, and support the development of STEM expertise that is epistemically robust, ethically grounded, and responsive to the complexities of real-world practice.

1.3 Argumentation Across STEM Disciplines and Professional Practice

Argumentation has been extensively documented as central to learning and professional practice across STEM disciplines. In science education, argumentation supports evidence-based reasoning, engagement with uncertainty, and understanding of the nature of science (Erduran et al., 2015; Koichu et al., 2022; Ogunniyi & Iwuanyanwu, 2024; Walker & Sampson, 2013). In mathematics, it underpins justification, proof, and generalisation. In engineering and technology, argumentation mediates design decisions, risk analysis, optimisation, and ethical trade-offs (Dym et al., 2005; Madhavan, 2015). Beyond formal education, professional practice relies on argumentation to coordinate collaborative work, justify decisions to stakeholders, and manage the consequences of innovation. Studies of professional reasoning highlight how engineers, scientists, and technologists routinely engage in argumentation to negotiate constraints, evaluate alternatives, and defend courses of action in high-stakes contexts (Gainsburg et al., 2016; Gouran et al., 1994).

Figure 1.4 visually represents disciplinary variations of argumentation across STEM through four overlapping domains, science, mathematics, engineering, and technology, with mathematics positioned centrally. This central placement signifies mathematics as a shared epistemic resource that undergirds reasoning practices across the other STEM disciplines. The overlaps illustrate shared epistemic

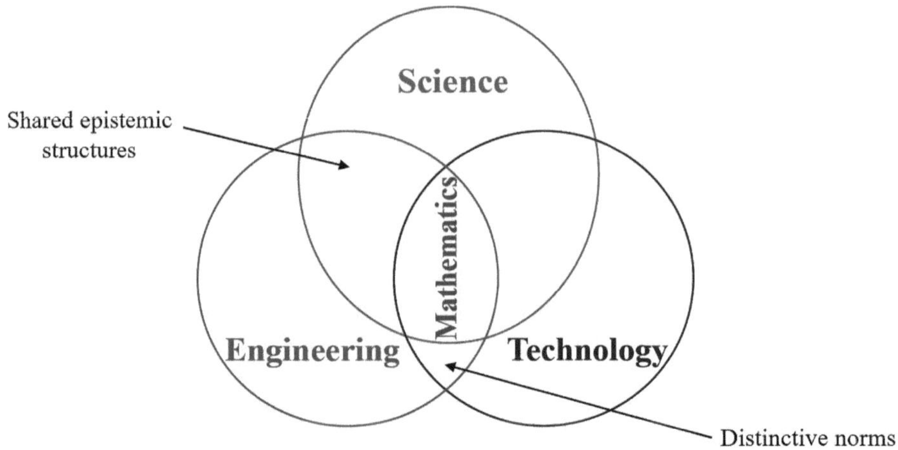

Fig. 1.4 Disciplinary variations of argumentation

structures, such as the use of evidence, logical reasoning, modelling, and justification, while the outer, less-overlapping regions highlight distinctive norms that characterise argumentation within each discipline.

In science, the diagram emphasises argumentation oriented toward empirical evidence, uncertainty, and theory–data coordination. This aligns with the earlier discussion on arguing to learn across STEM practices, where argumentation supports evidence-based reasoning and engagement with the nature of scientific knowledge. In mathematics, argumentation is shown as more tightly structured and abstract, focusing on justification, proof, and generalisation, reflecting its normative emphasis on deductive validity rather than empirical warrant. The engineering and technology domains overlap substantially with science and mathematics but extend outward to represent their distinctive emphasis on design-oriented argumentation, where decisions are negotiated through trade-offs among constraints, optimisation criteria, risk, feasibility, and ethical considerations. Read alongside the text in Table 1.1, the diagram underscores that while argumentation is a unifying practice across STEM, it is not epistemically uniform. Instead, it is shaped by the purposes, methods, and accountability structures of each discipline. This directly supports the claim that professional STEM practice relies on argumentation to justify decisions, manage uncertainty, and coordinate collaborative work in high-stakes contexts.

1.4 A Practitioner-Facing Orientation

This handbook is written for STEM educators, researchers, professional practitioners, and students preparing to enter STEM professions. Our orientation is explicitly practitioner-facing. We do not treat argumentation as an abstract pedagogical

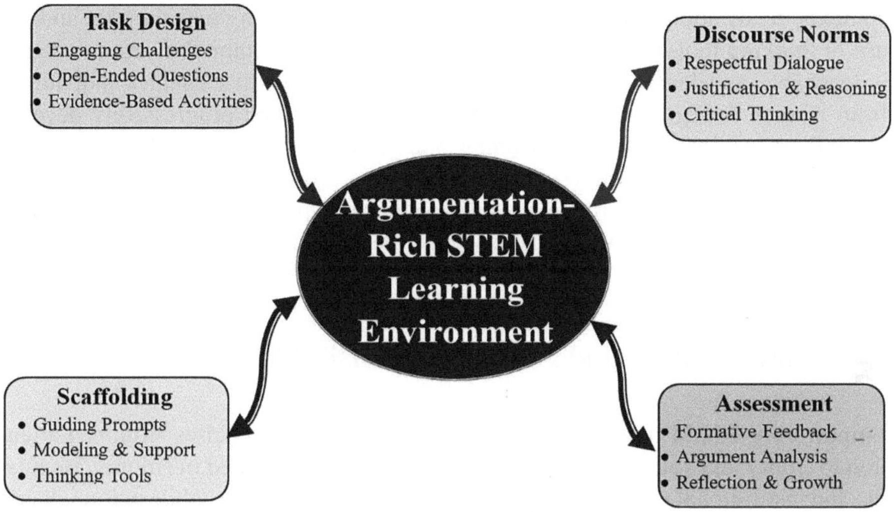

Fig. 1.5 Design principles for argumentation-rich STEM learning environments

ideal, but as a practical, teachable, and assessable form of professional reasoning. Throughout the volume, we foreground:

(i) authentic tasks drawn from scientific inquiry, engineering design, and technological problem-solving;
(ii) design principles for embedding argumentation within curricula and professional learning environments;
(iii) analytic frameworks for evaluating the quality of arguments;
(iv) strategies for scaffolding participation, feedback, and assessment;
(v) considerations of equity, ethics, and responsibility in argument-rich contexts.

Figure 1.5 illustrates a practice-oriented set of design principles for STEM learning environments in which argumentation is deliberately cultivated through the coordinated interaction of four core components: task design, discourse norms, scaffolding, and assessment. Rather than presenting argumentation as a standalone construct, the diagram emphasises how these components work together to embed professional reasoning into everyday teaching, learning, and practice, consistent with the explicitly practitioner-facing orientation. Task design foregrounds engaging challenges, open-ended questions, and evidence-based activities. This reflects the handbook's commitment to authentic STEM tasks drawn from scientific inquiry, engineering design, and technological problem-solving, where students and practitioners must weigh evidence, justify choices, and respond to uncertainty and constraints.

Discourse norms highlight respectful dialogue, justification and reasoning, and critical thinking. These norms establish the social and ethical conditions under which productive argumentation can occur, aligning with the emphasis on professional communication, accountability, and responsible participation in STEM contexts. Scaffolding focuses on guiding prompts, modelling and support, and the use of thinking tools. This component operationalises the claim that argumentation is teachable and learnable, providing structured support that enables individuals from diverse backgrounds to participate meaningfully and progressively develop disciplinary and professional reasoning practices. Assessment emphasises formative feedback, argument analysis, and reflection for growth. This reinforces the view that argumentation is assessable and improvable, and that evaluation should support learning trajectories rather than merely judge performance.

1.5 Equity, Voice, and Ethical Responsibility

Empowering STEM thinkers through argumentation also requires sustained attention to equity and voice. Argumentation practices are embedded in cultural, linguistic, and institutional norms that may privilege some participants while marginalising others. Research demonstrates that explicit instruction, shared norms, and structured scaffolds are essential for ensuring broad and equitable participation in argumentative discourse (Erduran et al., 2004; Vellom & Anderson, 1999). We approach argumentation as both a cognitive and a sociocultural practice, framing STEM learning within the broader contexts necessary to challenge hegemonic STEM traditions (Zeidler, 2016). Supporting equitable participation involves valuing diverse forms of evidence, legitimising multiple ways of knowing, and positioning critique as a collective epistemic responsibility rather than as personal contestation. For STEM practitioners, this is not only a pedagogical concern but also an ethical one, directly linked to responsible innovation and social accountability (Evagorou & Osborne, 2013; Sadler, 2006).

1.6 Overview of the Book Chapters

The chapters in this book are deliberately sequenced to move from foundational theory to disciplinary application and professional practice. Chapters 1 and 2 introduce the purpose, scope, and educational rationale for emphasising argumentation in STEM as a pathway to cultivating critical, ethical, and future-oriented thinkers. It situates argumentation as a core human cognitive skill required to navigate the epistemic, ethical, and decision-making challenges posed by frontier technologies such as artificial intelligence, CRISPR gene editing, autonomous robotics, and interconnected IoT systems—technologies that increasingly shape how knowledge is generated, validated, and applied in professional STEM contexts.

Chapter 3 examines what arguments are, and are not, clarifying core concepts, structures, and norms that undergird reasoned judgment across STEM disciplines. Chapter 4 develops the epistemic practices required for constructing and evaluating arguments, with attention to evidence, justification, and critique in complex and uncertain problem spaces. Chapter 5 situates argumentation within the mind–machine nexus of contemporary STEM, examining how human reasoning interacts with computational models, automated systems, and data-driven technologies. Chapters 6 and 7 examine scientific practices and engineering design thinking as argument-rich enterprises in which practitioners must negotiate trade-offs, risks, and values. Chapter 8 extends argumentation into connectivist and digitally networked learning environments, reflecting the distributed and collaborative nature of knowledge work in an increasingly interconnected technological landscape. Chapter 9 synthesizes the book's themes by demonstrating how argumentation underpins STEM learning, epistemic apprenticeship, critical praxis, professional reasoning, reflective judgement and ethical action across diverse STEM contexts.

1.7 A Note to All Readers

As we embark on this inquiry together, I invite you to reconceptualise argumentation not as an added burden on already crowded curricula or professional routines, but as a unifying cognitive and epistemic practice that makes STEM thinking visible, accountable, and improvable. Across science, technology, engineering, and mathematics, argumentation is the means by which knowledge is not merely transmitted, but examined, justified, challenged, and refined. By centring argumentation, we strengthen the intellectual integrity of STEM education and professional practice, and we empower learners and practitioners alike to participate responsibly in shaping scientific and technological futures. In an era increasingly shaped by artificial intelligence, automation, and data-driven systems, the distinctive value of human cognition lies not in speed or computational capacity, but in judgment, reasoning, and ethical discernment. Machines may calculate faster and process vast amounts of information, but they cannot replace the human capacity to question assumptions, weigh evidence, or deliberate about consequences. These capacities are cultivated through argumentation. Engaging in structured, evidence-based argumentation enables us to navigate uncertainty, complexity, and disagreement, conditions that now define much of contemporary STEM work.

It is crucial, however, to clarify what argumentation is, and what it is not. Argumentation in STEM is not about winning debates, overpowering others, or defending positions at all costs. When argument is framed as a contest to be won, inquiry is narrowed and intellectual humility is lost. Scientific and technological progress depends not on always being right, but on remaining open to revision in light of stronger reasoning or new evidence. Argumentation, as it is treated in this book, is a collaborative and ethical practice oriented toward understanding, improvement, and shared knowledge-building rather than personal victory. Throughout this book, argumentation is presented as the architecture of thought itself, the structure through which ideas are generated, tested, and refined. Analytical, dialectical, and rhetorical forms of argumentation are examined as complementary dimensions of reasoning that together support rigorous STEM practice. These forms equip students, practitioners and emerging professionals to evaluate claims, challenge assumptions, communicate persuasively, and make reasoned decisions in contexts where the stakes are high and certainty is rare.

Although I write from a particular perspective, this work is offered in a spirit of inquiry rather than finality. I invite you to engage with the arguments presented here as an active intellectual partner; to question, critique, and extend them within your own disciplinary and professional contexts. In doing so, we participate together in the very practice this book seeks to foreground, argumentation as a means of thinking deeply, learning continuously, and acting wisely in a rapidly changing technological world. Without argumentation, STEM risks becoming technically proficient yet intellectually shallow and ethically fragile. With argumentation, STEM retains coherence, creativity, resilience, and responsibility. We do not merely learn what STEM knowledge is; we learn how it is constructed, evaluated, and revised through empirical investigation and logical justification. We develop the capacity to function as critical friends, reflective practitioners, and responsible contributors to scientific progress and societal problem-solving. Ultimately, argumentation stands as a transformative pedagogical and professional commitment. It fosters deeper understanding, disciplined inquiry, and lifelong intellectual growth. More than a skill, it is the disciplined practice through which reasoning matures into judgment, knowledge evolves into insight, and technical expertise develops into wisdom.

References

Asterhan, C. S. C., & Schwarz, B. B. (2007). The effects of monological and dialogical argumentation on concept learning in evolutionary theory. *Journal of Educational Psychology, 99*(3), 626–639. https://doi.org/10.1037/0022-0663.99.3.626

Berland, L. K., & McNeill, K. L. (2010). A learning progression for scientific argumentation: Understanding student work and designing supportive instructional contexts. *Science Education, 94*(5), 765–793. https://doi.org/10.1002/sce.20402

Driver, R., Newton, P., & Osborne, J. (2000). Establishing the norms of scientific argumentation in classrooms. *Science Education, 84*(3), 287–312. https://doi.org/10.1002/(SICI)1098-237X(200005)84:3<287::AID-SCE1>3.0.CO;2-A

Duschl, R. A., & Osborne, J. (2002). Supporting and promoting argumentation discourse in science education. *Studies in Science Education, 38*(1), 39–72. https://doi.org/10.1080/03057260208560187

Dym, C. L., Agogino, A. M., Eris, O., Frey, D. D., & Leifer, L. J. (2005). Engineering design thinking, teaching, and learning. *Journal of Engineering Education, 94*(1), 103–120. https://doi.org/10.1002/j.2168-9830.2005.tb00832.x

Erduran, S., Ozdem, Y. & Park, J. Y. (2015). Research trends on argumentation in science education: a journal content analysis from 1998–2014. *International Journal of STEM Education, 2*(5), 1–12. https://doi.org/10.1186/s40594-015-0020-1

Erduran, S., Simon, S., & Osborne, J. (2004). Tapping into argumentation: Developments in the application of Toulmin's argument pattern for studying science discourse. *Science Education, 88*(6), 915–933. https://doi.org/10.1002/sce.20012

Evagorou, M., & Osborne, J. (2013). Exploring young students' collaborative argumentation within a socio-scientific issue. *Journal of Research in Science Teaching, 50*(2), 209–237. https://doi.org/10.1002/tea.21076

Gainsburg, J., Fox, J., & Solan, L. M. (2016). Argumentation and decision making in professional practice. *Theory Into Practice, 55*(4), 332–341. https://doi.org/10.1080/00405841.2016.1208072

Gee, J. (2015). *Social linguistics and literacies: Ideology in discourses* (5th ed.). https://doi.org/10.4324/9781315722511

Gouran, D. S., Miller, L. D., Wiethoff, W. E., & Doelger, J. A. (1994). Mastering communication (2nd ed.). Allyn and Bacon.

Iwuanyanwu, P. N. (2020). Nature of problem-solving skills for 21st Century STEM Learners: What teachers need to know, *Journal of STEM Teacher Education, 55*(1), 27–40. https://doi.org/10.30707/JSTE55.1/MMDZ8325

Jiménez-Aleixandre, M. P., & Erduran, S. (2007). Argumentation in science education: An overview. In S. Erduran & M. P. Jiménez-Aleixandre (Eds.), *Argumentation in science education* (Science & technology education library) (Vol. 35). Springer. https://doi.org/10.1007/978-1-4020-6670-2_1

Kelly, G. J., Regev, J., Prothero, W. (2007). Analysis of lines of reasoning in written argumentation. In: Erduran, S., Jiménez-Aleixandre, M.P. (eds) Argumentation in science education. *Science & Technology Education Library*, (Vol. 35) Springer, Dordrecht. https://doi.org/10.1007/978-1-4020-6670-2_7

Koichu, B., Schwarz, B. B., Heyd-Metzuyanim, E., Tabach, M., & Yarden, A. (2022). Design practices and principles for promoting dialogic argumentation via interdisciplinarity. *Learning, Culture and Social Interaction*, 37:100657. https://doi.org/10.1016/j.lcsi.2022.100657

Madhavan, G. (2015). *Applied minds: How engineers think.* W. W. Norton & Company.

Nussbaum, E. M. (2011). Argumentation, dialogue theory, and probability modeling: Alternative frameworks for argumentation research in education. *Educational Psychologist, 46*(2), 84–106. https://doi.org/10.1080/00461520.2011.558816

Ogunniyi, M. B. (2007). Teachers' stances and practical arguments Regarding a Science-indigenous Knowledge curriculum: Part 2. *International Journal of Science Education, 29*(10), 1189–1207. https://doi.org/10.1080/09500690600931038

Ogunniyi, M., & Iwuanyanwu, P. N. (2024). Analysis of Teachers' perspectives towards the use of IKS to Improve STEM Education for sustainable development. *African Journal of Research in Mathematics, Science and Technology Education, 28*(3), 319–329. https://doi.org/10.1080/18117295.2024.2352980

Osborne, J. (2010). Arguing to learn in science: The role of collaborative, critical discourse. *Science, 328*(5977), 463–466. https://doi.org/10.1126/science.1183944

Osborne, J., Erduran, S., & Simon, S. (2004). Enhancing the quality of argumentation in school science. *Journal of Research in Science Teaching, 41*(10), 994–1020. https://doi.org/10.1002/tea.20035

Ryu, S., & Sandoval, W. A. (2012). Improvements to elementary children's epistemic understanding from sustained argumentation. *Science Education, 96*(3), 488–526. https://doi.org/10.1002/sce.21006

Sadler, T. D. (2006). Promoting discourse and argumentation in science teacher education. *Journal of Science Teacher Education, 17*, 323–346. https://doi.org/10.1007/s10972-006-9025-4

Sampson, V., & Clark, D. B. (2008). Assessment of the ways students generate arguments in science education: Current perspectives and recommendations for future directions. *Science Education, 92*(3), 447–472. https://doi.org/10.1002/sce.20276

Toulmin, S. (1958/2003). *The uses of argument* (Updated ed.). Cambridge University Press.

Vellom, R. P., & Anderson, C. W. (1999). Reasoning about data in middle school science. *Journal of Research in Science Teaching, 36*(2), 179–199.

Walker, J. P., & Sampson, V. (2013). Learning to argue and arguing to learn: Argument-driven inquiry as a way to help undergraduate chemistry students learn how to construct arguments and engage in argumentation during a laboratory course. *Journal of Research in Science Teaching, 50*(5), 561–596. https://doi.org/10.1002/tea.21082

Zeidler, D. L. (2016). STEM education: A deficit framework for the twenty first century? A sociocultural socioscientific response. *Cultural Studies of Science Education, 11*(2016), 11–26. https://doi.org/10.1007/s11422-014-9578-z

Chapter 2
Adapting Human Cognitive Skills in the Twenty-First-Century Industrial Revolutions

2.1 Introduction

Frontier technologies, including artificial intelligence (AI), automation, advanced robotics, genetic engineering through clustered regularly interspaced short palindromic repeats (CRISPR), and the Internet of Things (IoT), were once positioned at the margins of STEM practice. Today, however, they occupy its centre. They define its operational terrain, shape its epistemic priorities, and direct its future innovation. Consequently, they are actively reshaping the cognitive ecology within which scientific inquiry, engineering design, and technological decision-making occur. Across laboratories, classrooms, industrial systems, and public infrastructures, we increasingly work within socio-technical arrangements that think with us, act alongside us, and, in some cases, recommend or even pre-empt professional judgments. Importantly, this shift aligns with socio-technical and biosocial-technical perspectives, which emphasise that technology's effects are never purely technical but are shaped by organisational choices, policy contexts, and human agency (Dankbaar, 1988; Sony & Naik, 2020; Walker et al., 2008).

In earlier industrial revolutions, human cognition was positioned as the central engine of production and problem-solving. Professional competence was largely equated with procedural mastery, technical accuracy, and the reliable execution of predefined methods. Even as tools became more sophisticated, the human expert remained the primary locus of interpretation and judgment. By contrast, in the current industrial transformations associated with Industry 4.0, this assumption no longer holds (Schwab & Davis, 2018). Intelligent systems now execute many forms of routine cognitive labour, classification, optimisation, prediction, and control, with speed, scale, and consistency that exceed human capability. Moreover, contemporary analyses of AI, automation, and robotics demonstrate that these systems increasingly shape not only how tasks are performed but also how problems are framed, prioritised, and evaluated (Frey & Osborne, 2017; OECD, 2019).

P. N. Iwuanyanwu, *Empowering STEM Thinkers Through Argumentation*, Contemporary Trends and Issues in Science Education 69,
https://doi.org/10.1007/978-3-032-27906-4_2

In light of these developments, the central question for STEM professionals and those preparing to enter these fields cannot be reduced to technical upskilling alone. Rather, we must ask how forms of reasoning that remain meaningful, authoritative, and ethically grounded can be cultivated in environments saturated with machine intelligence. This concern, in turn, resonates strongly with policy and scholarly debates that caution against viewing AI adoption as technologically deterministic and instead highlight the need for deliberate human governance and responsibility (Krafft et al., 2020; Nguyen et al., 2023). Accordingly, we are compelled to confront a deeper question: *How do we, as STEM educators, researchers, practitioners, and emerging professionals, preserve and amplify what is distinctly human in an era shaped by intelligent machines?* This question is not merely technical; it is fundamentally philosophical and pedagogical. While intelligent systems excel at pattern recognition and optimisation within predefined parameters, they do not possess intrinsic values, moral accountability, or lived experience. Indeed, as numerous studies on AI and the future of work suggest, the real comparative advantage of humans lies not in computational speed but in judgment, creativity, ethical reasoning, and social interpretation (Anderson & Rainie, 2018; Wang & Siau, 2019). Therefore, our task is not to compete with machines on their own terms. Instead, it is to strengthen human capacities for critical judgment, creative reasoning, ethical discernment, and interpretive meaning-making. In doing so, we reposition STEM education and practice as sites not of technological substitution, but of deliberate human development within an increasingly intelligent technological world.

2.2 Why Cognition Must Change Now

Concerns regarding the displacement of human labour through automation have accompanied technological change for decades. Economic and sociological analyses have long warned that intelligent machines could displace skilled, semi-skilled, and unskilled workers alike, fundamentally altering professional structures and labour markets (Autor, 2015; Berg et al., 2018; Dankbaar, 1988). However, what has become increasingly apparent is that automation's influence extends beyond employment into cognition itself. Indeed, as documented in the AI–Automation–Robotics literature, intelligent systems reshape how evidence is produced, how alternatives are evaluated, and how decisions must be justified within professional and public domains (Brynjolfsson et al., 2018; Frank et al., 2019).

Consequently, algorithms now recommend medical diagnoses, rank engineering design options, optimise logistics, and inform policy-relevant modelling. Professional responsibility therefore expands to include the capacity to interrogate these systems, to understand their assumptions and limitations, and to account for their social, ethical, and equity implications (Latonero, 2018; Manjarrés et al., 2021; OECD, 2019). Moreover, this expansion of responsibility is particularly salient given evidence that AI-driven systems can exacerbate inequality, embed bias, and intensify surveillance if left unexamined (Karpouzis, 2024; Nazareno & Schiff, 2021).

From this perspective, the defining challenge for contemporary STEM practice is not whether humans can keep pace with machines, but whether we can cultivate cognitive capacities that machines cannot replicate. Critical thinking enables professionals to evaluate the credibility, bias, and contextual relevance of algorithmic outputs. Creative reasoning allows the imagination of alternative problem framings and futures beyond those optimised by existing systems. Ethical judgment anchors technical decision-making in considerations of safety, equity, sustainability, and public trust. Furthermore, research on worker well-being and automation suggests that these human capacities are central not only to professional effectiveness but also to meaning, autonomy, and long-term resilience in technologically mediated work environments (Davis et al., 2012; Gutek & Winter, 1990; Nazareno & Schiff, 2021).

Importantly, research within artificial intelligence itself reinforces this distinction. Advances in computational argumentation demonstrate that machines can participate in structured reasoning by generating and evaluating arguments within formalised frameworks (Bench-Capon & Dunne, 2007; Dung, 1995). These approaches have been extended into domains such as software engineering, design reasoning, and decision-support systems, thereby enhancing transparency and consistency in complex professional contexts (Demoly et al., 2019; Hästbacka & Kuikka, 2011). At the same time, these developments expose the limits of mechanised reasoning. Computational argumentation operates within predefined representations of claims, evidence, and warrants. It cannot autonomously determine which values ought to matter, whose interests should be prioritised, or how uncertainty should be ethically negotiated in novel or contested situations. Thus, as critical perspectives on AI governance emphasise, human judgment remains indispensable precisely because it is situated, contextual, and morally accountable (Krafft et al., 2020; Latonero, 2018).

In light of these considerations, the role of the STEM professional is shifting from executor to interpreter, critic, and steward of complex socio-technical systems. The enhancement of human cognitive capacities through argumentation is therefore not an optional enrichment to STEM practice; it is its intellectual lifeblood. Argumentation, as developed throughout this handbook, is not synonymous with debate or persuasion. Rather, it is the disciplined process through which practitioners link claims to evidence, evaluate competing explanations, justify decisions under uncertainty, and make their reasoning open to scrutiny.

As illustrated in Fig. 2.1, those who will thrive in contemporary STEM environments are not merely those who can operate advanced technologies, but those who can interrogate them. They ask how models are constructed, what assumptions are embedded in data, and which consequences follow from particular design or policy choices. Significantly, at the top of Fig. 2.1 we are positioned within a socio-technical ecology shaped by AI, data, and digital technologies. These are not depicted as peripheral tools but as structuring forces that feed directly into the cognitive processes through which learning, judgment, and decision-making now occur.

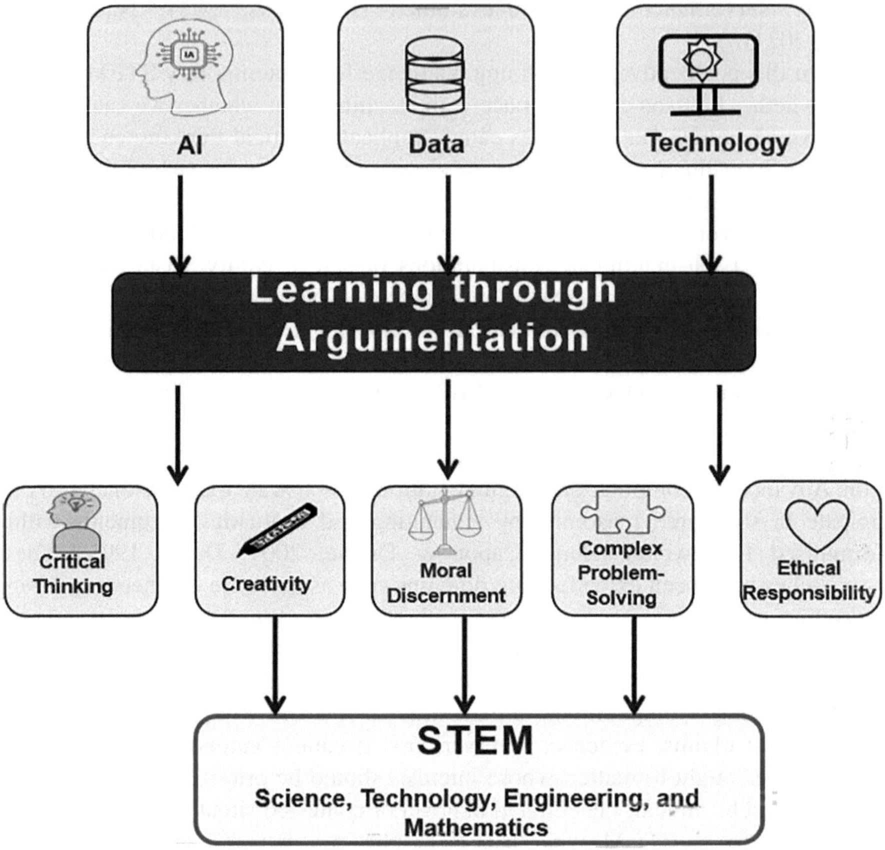

Fig. 2.1 Argumentation as the foundation of human judgment in AI-augmented STEM practice

In this way, the figure echoes our earlier claim that we no longer work with technology as an external aid, but within technological systems that actively participate in how problems are framed, evaluated, and resolved.

At the centre of the figure, learning through argumentation functions as the mediating layer between intelligent systems and human cognition. Crucially, argumentation is not treated as an add-on skill or a communicative afterthought. Instead, it is positioned as the core mechanism through which data, algorithmic outputs, and technological affordances are rendered meaningful, contestable, and ethically accountable. This central placement reflects the inquiry running throughout this introduction, namely how we preserve and strengthen forms of reasoning that remain distinctly human in environments saturated with machine intelligence. Positioned beneath argumentation, the figure foregrounds five interrelated human capacities that include critical thinking, creativity, moral discernment, complex problem-solving, and ethical responsibility. These are precisely the capacities that intelligent systems can support but cannot replace entirely. By visually aligning these capacities beneath argumentation, the figure makes visible our claim that such forms of reasoning are cultivated through disciplined engagement with evidence, claims, alternatives, and consequences, rather than through exposure to technology alone.

The grounding of these capacities within STEM underscores that this is not a philosophical abstraction removed from practice. Rather, the figure situates argumentation as the intellectual connective tissue linking science, technology, engineering, arts, and mathematics in contemporary professional contexts. We are thus led to the conclusion that thriving in Industry 4.0 does not depend solely on mastering frontier technologies, but on developing the judgment, creativity, and ethical imagination required to interrogate and steward them responsibly. In this light, Fig. 2.1 makes visible the chapter's central thesis, emphasizing that the future of STEM belongs not merely to those who can deploy intelligent systems, but to those who can reason with, question, and govern them wisely. The era of purely transactional STEM knowledge, where competence is defined by procedural fluency alone, has passed. What is now required is a deeper, more resilient form of cognition rooted in reflective practice, evidence-based reasoning, and ethical imagination, consistent with calls for education and training that foreground human-centred AI rather than AI-centred education (Karpouzis, 2024).

Within this broader educational and professional landscape, argumentation also performs a humanising function. It enables practitioners and future professionals to question the assumptions underlying AI models, evaluate the ethical boundaries of gene-editing technologies, scrutinise data generated by distributed IoT systems, and design robotic solutions that foreground human-centred values. These challenges cannot be delegated to machines. They demand interpretive sensitivity, moral awareness, and the capacity to reason across disciplinary, social, and ethical boundaries (OECD, 2019; Walker et al., 2008). By positioning argumentation at the core of this handbook, we seek to bridge two defining imperatives of contemporary STEM practice, technical fluency and human-centred judgment. Through sustained engagement with inquiry, critique, creativity, and ethical responsibility, we prepare practitioners and aspiring professionals to navigate the shifting frontiers of technological innovation with clarity, confidence, and purpose. Ultimately, the future of STEM depends not only on smarter machines, but on wiser humans, capable of reasoning, judging, and acting responsibly within complex and consequential socio-technical worlds.

2.3 The Risk of Surrendering Human Judgment to Intelligent Systems

As we trace the accelerating influence of frontier technologies, we must also confront an uncomfortable reality: human judgment is increasingly at risk in AI-mediated environments. Our relationship with intelligent systems is progressively characterised by a gradual, almost imperceptible surrender of cognitive agency. Across contemporary STEM practice, there is a growing tendency to outsource human judgment to digital machines, relying on artificial intelligence to compose texts, analyse data, optimise decisions, and generate recommendations that are often accepted with minimal scrutiny. This phenomenon aligns with concerns raised in policy and labour-market analyses, which caution that advanced automation risks repositioning humans from active epistemic agents to passive supervisors of machine outputs (OECD, 2019). For STEM practitioners, this shift presents a profound professional risk; for students preparing to enter these fields, it poses an even deeper educational one. When algorithms routinely perform cognitive work that was once central to professional reasoning, there is a danger that core analytical and evaluative capacities may weaken over time. Research on automation and skills use consistently shows that when cognitive tasks are displaced rather than augmented, workers experience skill erosion and reduced autonomy (Manjarres et al., 2021). The issue, therefore, is not the presence of intelligent assistance itself. Assistance has always been integral to scientific and technological progress. Rather, the concern lies in the subtle displacement of human judgment by automated inference and in the growing temptation to accept algorithmic outputs without interrogation.

This displacement carries significant consequences for professional practice. In STEM fields, disciplinary expertise has historically been grounded in disciplined skepticism, reasoned justification, and accountable decision-making. When practitioners no longer ask why a model behaves as it does, how an algorithm reaches its conclusions, or under what assumptions a result holds, professional expertise risks being reduced to procedural compliance rather than epistemic authority. Empirical studies of AI implementation in organisational settings reinforce this concern, warning that opaque systems can foster automation bias, diminish critical oversight, and weaken professional accountability (Nguyen et al., 2023). For this reason, adaptation to frontier technologies must extend beyond technical competence or operational fluency. It must also involve the deliberate preservation of intellectual agency, ethical responsibility, and philosophical self-awareness. These qualities are not abstract ideals but professional obligations, particularly in contexts where AI systems operate at scales and speeds that exceed human intuition.

As scholars of socio-technical systems argue, meaningful human oversight depends not merely on access to systems but on the capacity to interpret, question, and contest technological outputs (Walker et al., 2008). In this respect, argumentation functions as a cognitive safeguard, an anchoring epistemic practice that preserves human judgment within technologically saturated decision environments. At the same time, it is important to acknowledge the counterargument. Artificial intelligence, robotics, and automation offer unprecedented opportunities to enhance productivity, innovation, and economic growth. Empirical projections suggest that AI-driven technologies may contribute trillions of dollars to the global economy and generate new categories of work, particularly in data-intensive and creative domains (McKinsey Global Institute, 2017; World Economic Forum [WEF], 2024). From this perspective, delegating routine cognitive labour to machines enables practitioners to focus on higher-order tasks such as strategic reasoning, synthesis, and innovation.

However, this argument rests on a fragile assumption: namely, that higher-order thinking will persist automatically once lower-order reasoning is delegated. Historical and empirical evidence offers little support for this view. Studies of technological change consistently show that cognitive capacities not actively exercised tend to atrophy, particularly when work becomes increasingly system-directed rather than judgment-driven (Dankbaar, 1988; Davis et al., 2012). Thus, the promise of cognitive liberation through automation must be approached with critical caution rather than unqualified optimism. The deeper pedagogical challenge confronting contemporary STEM education, therefore, is not whether artificial intelligence should be integrated, but how it can be incorporated without undermining the development of judgment, discernment, and responsibility. As Nguyen et al. (2023) argue, education must prioritise AI literacy rather than mere AI usage, ensuring that learners understand the limits, assumptions, and ethical implications of intelligent systems. Accordingly, STEM graduates must become more than technically competent operators of advanced systems; they must be cognitively vigilant and ethically grounded professionals.

This challenge extends beyond individual cognition to questions of professional identity and responsibility. Across global contexts, practitioners increasingly struggle to locate their agency, accountability, and purpose within AI-driven economies. Labour-market analyses highlight growing concerns about accountability, bias, and the erosion of meaningful work as automation intensifies (Frey & Osborne, 2017; Korinek & Stiglitz, 2018). In this context, STEM education cannot be reduced to employability training alone. Rather, it must cultivate intellectual citizenship, the capacity to participate responsibly in knowledge production, technological governance, and public reasoning about science and technology. Intellectual citizenship requires that practitioners retain authorship of judgment. A professional who cannot explain or defend a decision without deferring to an algorithm has effectively abdicated epistemic and ethical responsibility. This concern is echoed in AI governance literature, which emphasises that accountability cannot be delegated to machines without undermining human responsibility (Latonero, 2018).

Philosophical traditions further deepen this argument by clarifying the relationship between reasoning and autonomy. From Kant's claim that reason constitutes the foundation of autonomy to Descartes's insistence that doubt initiates knowledge, the history of philosophy affirms that thinking is inseparable from responsibility. In algorithmically mediated environments, where system logics are often proprietary and opaque, this insight becomes especially urgent. Without disciplined skepticism, practitioners risk mistaking probabilistic outputs for epistemic certainty, a danger increasingly recognised in critical studies of AI ethics and governance (Greene et al., 2019; Nguyen et al., 2023). To think critically in STEM, therefore, is to move beyond procedural problem-solving into ethical and philosophical inquiry. Every algorithmic system redistributes power, risk, and opportunity; consequently, questions of fairness, accountability, and social impact become intrinsic to professional practice rather than peripheral concerns.

Aristotle's concept of *phronesis*, or practical wisdom, reinforces this point by reminding us that technical knowledge achieves its full value only when exercised with judgment attuned to context and consequence. In light of this, and amid the accelerating pace of technological change, the future will favour not merely those who possess the most knowledge, but those who are capable of thinking with depth and responsibility. Reflection, moral reasoning, and intellectual accountability remain irreducibly human capacities. For STEM practitioners, and for those preparing to enter these professions, the task is therefore not simply to adapt to intelligent machines, but to engage them critically, supervise them responsibly, and shape them through ethically grounded and philosophically informed judgment. This, we contend, is the enduring purpose of STEM education in the twenty-first century.

2.4 Advancements in Frontier Technologies

Building on the preceding discussion, as we observe the rapid expansion of AI, automation, and robotics, we cannot ignore how profoundly these frontier technologies are reshaping the landscape of STEM education and professional practice. What is commonly described as the Fourth Industrial Revolution represents not a single technological breakthrough, but rather a convergence of advances in computing power, data availability, machine learning, and intelligent automation that increasingly simulate or replace cognitive tasks once considered uniquely human (Frey & Osborne, 2017; Schwab & Davis, 2018). This convergence, in turn, has dissolved long-standing boundaries between human cognition and machine intelligence, compelling us to rethink what it means to learn, to reason, and to know in technologically mediated environments. Consequently, STEM education assumes a position of unparalleled importance. National and international policy analyses consistently emphasize that education systems must move beyond training students to merely operate advanced technologies and instead prepare them to understand their underlying logics, limitations, and consequences (Karpouzis, 2024; OECD, 2019).

At the core of this educational imperative lie critical inquiry and argumentation, intellectual practices we regard as foundational not only to STEM disciplines but to education and civilization itself. Without these practices, there is a real danger of producing graduates who are technically proficient yet epistemically dependent, individuals capable of deploying powerful systems without the capacity to question their assumptions or societal implications (Nazareno & Schiff, 2021). In this sense, critical inquiry is not merely a methodological technique but an orientation of mind. It cultivates the disposition to ask questions that matter, to probe beneath surface-level explanations, and to challenge the taken-for-granted authority often attributed to algorithmic outputs. This orientation becomes especially vital in AI-driven contexts, where systems are frequently perceived as neutral or objective despite being trained on historically situated and socially biased data (Nguyen et al., 2023). Through inquiry, learners are empowered to interrogate data rather than accept it uncritically, to recognize bias embedded in models, and to understand that technological systems are socio-technical constructions rather than value-free tools (Sony & Naik, 2020).

From a philosophical standpoint, this position challenges instrumentalist views of technology that prioritize efficiency, optimization, and predictive accuracy as primary educational goals. Proponents of such views argue that intelligent automation enhances productivity and reduces human error, particularly in complex STEM domains (Dong & Mclntyre, 2014). While we acknowledge these gains, the literature also cautions that overreliance on automated decision systems can erode human judgment and deskill professional reasoning if critical engagement is absent (Berg et al., 2018). Inquiry, therefore, functions as a safeguard against epistemic complacency, cultivating what we describe as intellectual courage, the willingness to remain with uncertainty long enough for understanding to deepen. Closely related to inquiry, argumentation serves as its indispensable dialogical counterpart. Through argumentation, claims are tested, evidence is scrutinized, and reasoning is rendered transparent and accountable (Iwuanyanwu, 2020).

Research on AI-integrated workplaces suggests that the most resilient professionals are not those who compete with machines, but those who can justify decisions, evaluate competing outputs, and communicate reasoning across human–machine interfaces (Frank et al., 2019). Accordingly, when students learn to construct claims, support them with evidence, and respond thoughtfully to counterarguments, they develop epistemic integrity rather than procedural compliance. In an era marked by information saturation and algorithmically curated knowledge, argumentation restores ownership of thought. It enables learners to resist the passive consumption of machine-generated outputs and to reclaim their role as active knowers and decision-makers. This capacity is particularly indispensable in STEM professions where AI increasingly shapes scientific discovery, engineering design, and medical diagnostics (Brynjolfsson et al., 2018).

For this reason, we have long argued that STEM education must move beyond rote memorization toward inquiry-driven learning that mirrors the authentic practices of scientific and technological work. For example, consider a biology classroom in which students debate the ethical implications of gene-editing technologies such as CRISPR. Such engagements reflect real scientific discourse, where evidence, uncertainty, and values intersect. In this way, the classroom becomes a microcosm of professional STEM practice, a space where knowledge is negotiated rather than transmitted (Leontief, 1974). Moreover, as technological innovation accelerates, the integration of advanced tools into STEM education is no longer optional; it is essential. AI-driven simulations, data analytics platforms, and virtual environments offer unprecedented opportunities to deepen inquiry and enrich argumentation.

When thoughtfully employed, these tools function both as instruments of investigation and as objects of critical reflection. Students may use AI to model climate systems or analyse complex datasets while simultaneously interrogating the ethical and epistemic implications of algorithmic mediation (Karpouzis, 2024). In particular, simulations can transform learning by rendering complex and otherwise inaccessible processes visible. As students manipulate variables, test hypotheses, and evaluate outcomes, they develop conceptual understanding alongside epistemic humility, an appreciation that scientific knowledge is provisional, model-dependent, and open to revision (Autor, 2015). As a result, argumentation is strengthened, since students must justify conclusions through evidence and reasoning rather than defer to authority.

For educators, however, this shift requires a willingness to relinquish control and to embrace uncertainty. Inquiry-driven teaching challenges deeply ingrained instructional habits by prioritizing questions over answers and disagreement over premature consensus. Nevertheless, this pedagogical orientation aligns with the demands of a world defined by volatility, complexity, and interdisciplinary problem-solving, conditions intensified by automation and intelligent systems (Nazareno & Schiff, 2021). In such a world, adaptability emerges as both a moral and intellectual necessity. The most valuable graduates will not be those who merely accumulate technical knowledge, but those who can think critically, argue persuasively, and act responsibly.

Design-based STEM projects, for instance, such as developing sustainable energy solutions or applying AI to healthcare, demand precisely these capacities by integrating technical expertise with ethical sensitivity and collaborative reasoning. Ultimately, the true purpose of STEM education in the era of intelligent machines is to cultivate minds that remain intellectually alive to complexity. By grounding inquiry in argumentation and ethical awareness, STEM education ensures that scientific and technological progress is guided by wisdom rather than driven by speed alone. It is within this essential balance, between automation and humanity, innovation and reflection, that STEM education fulfils its highest calling.

2.5 The Future of Work and Frontier Technologies' Impact on Jobs

As we look toward the future, a profound question arises: *will there still be a meaningful place for human labour in a world increasingly dominated by machines and digital technologies?* With the rapid advancements in artificial intelligence, robotics, and engineering, machines are now capable of performing a vast array of tasks that were once the exclusive domain of human effort (see Fig. 2.2). By 2050, we can anticipate significant transformations in how we work, communicate, and coexist within an increasingly automated world. Yet, amid these technological developments, it is crucial to remember that no AI system or robot, no matter how advanced, possesses consciousness. Robots and AI lack the creativity, spontaneity, imagination, and emotional depth that characterize human experience. They do not laugh at humour, appreciate art, or feel pain, qualities that define what it means to be human (Zohar & Marshall, 1990). While machines can simulate certain human behaviours or mimic affective responses, there remains a profound distinction between programmed performance and genuine consciousness (Minda, 2019). Reducing

Fig. 2.2 Automation tend to eliminate more jobs than it creates

ourselves to mere mechanical entities risks diminishing the richness of human existence, narrowing our understanding of intelligence and value to what can be encoded as algorithmic behaviour (Zohar & Marshall, 1990). If we are not vigilant, we may become desensitized to our own human experience, an alienation that could distort how we perceive ourselves, our consciousness, and our role in the world.

In this context, one must ask: (1) what does the future of human labour look like in an age of accelerating automation? (2) Are today's STEM students, tomorrow's scientists, engineers, and technologists, prepared to adapt to the fast-paced dynamics of the twenty-first-century industrial revolutions, where machines increasingly replace human roles in the workforce? These questions are not speculative; they determine how society will navigate the ethical and economic realities of automation. Proponents of automation often argue that, despite the displacement of both skilled and unskilled workers, automation holds the potential to accelerate progress toward a more equitable and dignified global society (WEF, 2017). From this perspective, AI technologies could foster cleaner production systems, safer environments, and enhanced human well-being (UN Economic and Social Council [UNECOSOC], 2017). Yet optimism alone cannot mask the profound social and economic challenges automation presents. The ideal of human–machine symbiosis, blending human intellect with machine precision, may yield productivity gains, but it also demands vigilance and ethical foresight.

We must acknowledge that technological innovation, while transformative, is also disruptive. The reality is that many jobs, both routine and complex, are being lost to automation. Early research by Levin and Rumberger (1987) predicted that ongoing technological progress would likely displace more jobs than it creates. That prediction has, in many ways, become our lived experience. From manufacturing to service industries, automation has already restructured employment patterns across

Fig. 2.3 The evolution from human to machine labour

both developed and developing nations. Although international organizations such as the WEF and UNECOSOC advocate for "reskilling" and "lifelong learning," these efforts represent only partial solutions. The pace of change outstrips the pace of adaptation. The scale of potential disruption is staggering. Estimates suggest that as many as 80% of existing jobs could be affected by automation in the coming decades (UNECOSOC, 2017). While economic growth and innovation may generate new roles in emerging industries, the transition will likely be turbulent and uneven (Leontief & Duchin, 1986). For workers trained in traditional trades or narrowly defined skills, the question becomes existential: *will there still be a place for them in this new, automated world?*

The social implications of this transformation are immense. Job losses have already accelerated in many advanced economies and are now increasingly evident in developing nations, where automation threatens both industrial and service sectors (see Figs. 2.3 and 2.4). The machines and algorithms humanity has created to serve progress now risk becoming instruments of alienation, a phenomenon some might call a human invasion. According to the United Nations Department of Economic and Social Affairs (UNDESA, 2019), the most pervasive global fear surrounding AI is its potential to replace human workers on a massive scale, resulting in widespread unemployment and social instability. The threat is particularly acute for low- and medium-skilled workers, whose manual and cognitive tasks are increasingly performed by sophisticated machines and AI-driven systems (UNECOSOC, 2017).

This growing asymmetry between technological capability and human opportunity represents one of the greatest moral and economic challenges of our time. To confront it, societies must prioritize education systems that cultivate adaptability, creativity, and ethical awareness, qualities that machines cannot replicate. As educators, policymakers, and learners, we must resist technological determinism, the belief that technological progress is an autonomous, unstoppable force. Instead, we should assert human agency over technology's direction and purpose. The future of work must not be defined by obsolescence but by reinvention, an ongoing commitment to ensure that AI and automation augment human potential rather than diminish it. Ultimately, the question is not whether machines will replace humans, but whether humans will redefine work itself in ways that preserve dignity, purpose, and intellectual freedom. The task before us is to envision a world where technology serves humanity, not the reverse, and to prepare generations capable of steering that vision with ethical conviction, creativity, and wisdom.

2.6 A New World Is Emerging in the Workplace

There is no longer any doubt that a new world of work is emerging, one that is fundamentally shaped by artificial intelligence (AI), automation, and the rise of autonomous machines. We are witnessing a gradual but profound reconfiguration of the boundary between human and robotic labour. What was once seen as collaboration, robots serving as mere "assistants" to human workers, is rapidly evolving into replacement. The steady encroachment of machines into domains once reserved for human expertise compels us to ask urgent questions about the future of human work and dignity. The growing dominance of sophisticated technologies and autonomous systems represents not only an economic challenge but also a human and educational crisis.

An entire generation of students risks entering a world where they may be outpaced, or even governed, by the very technologies they were trained to design. The anxiety surrounding AI's impact on employment is no longer speculative; it is becoming a lived experience. Studies predict that within the next two decades, up to 47% of jobs in the United States and 54% across the European Union could be automated (Bowles, 2017). In developing economies, the impact may be even more severe, with as much as 66% of work susceptible to automation (Frey & Osborne, 2017). Such transformations could erode middle-skilled professions, exacerbate income inequality, and destabilize labour markets on a global scale (Gries & Naudé, 2018).

As automation deepens its reach, we must confront a difficult truth: these changes may not herald universal prosperity but instead a new hierarchy of work, one defined by access to digital capital and algorithmic literacy. If the current trajectory continues unchecked, billions of people could find themselves excluded from meaningful economic participation. The convenience of self-service technologies, for instance, conceals a troubling undercurrent. When retail cashiers are replaced by automated checkouts, efficiency may rise, but human connection and employment decline. This is not an isolated phenomenon; it is a continuation of the industrial logic that began in the eighteenth century, extending what Marx once described as the alienation of labour.

As Zohar (1995) observes, personal and social relationships, those moral and emotional bonds that define our humanity, are increasingly detached from the mechanized, bureaucratic systems that dominate modern work. In this sense, automation is not merely transforming labour; it is reshaping the moral and social architecture of the workplace itself. Consider, for example, the rapid development of self-driving vehicles. While these innovations promise safety and efficiency, they also raise pressing ethical and economic questions:

(i) Are we celebrating only the technological triumphs and economic advantages that benefit a privileged few?
(ii) Who stands to gain, and who will be left behind?
(iii) What becomes of the millions of professional drivers whose livelihoods depend on this form of labour?
(iv) Should there be moral limits to how far we allow machines to replace human participation in work and society?

In previous industrial revolutions, such as the era of microelectronics, automation displaced certain types of physical labour, but new forms of employment eventually emerged to absorb the displaced workforce (Rumberger, 1984). This time, however, the scope and pace of technological change are unprecedented. The very systems we once designed to assist us now have the capacity to operate autonomously, learn from data, and make decisions without human oversight. This shift challenges the very notion of what it means to work. The social fabric of the workplace is already evolving. Soon, human-to-human collaboration may be replaced, or at least mediated, by human–robot interaction, as illustrated in Fig. 2.3. One can imagine a near future where the workplace dynamic transitions from 'human versus human' competition to 'human versus machine', and eventually, to 'machine versus machine', with humans observing from the periphery. The psychological and social consequences of such a shift could be profound.

Emerging technologies now span the digital, physical, and biological realms, intertwining to produce an era of unprecedented transformation. These include the Internet of Things (IoT), autonomous vehicles, augmented and virtual reality (AR and VR), and biotechnological innovations such as gene editing and nanotechnology. As biological engineering accelerates, it is conceivable that in a few decades, humans may no longer recognize themselves in their original form. The boundary between human and machine, once distinct, may blur to the point of indistinguishability. Even now, we are beginning to experience the emotional simulation of machines. Robots capable of mimicking affection and empathy are being integrated into domestic and social spaces. They greet us at home, offer companionship, and imitate care, behaviours once considered the exclusive domain of human intimacy.

While such developments may provide comfort or novelty, they also invite reflection on what we risk losing: the authenticity of human connection, the spontaneity of emotion, and the depth of moral responsibility that no algorithm can truly replicate. We are thus confronted with a paradox. The technologies designed to enhance human life also possess the potential to diminish its meaning. If we fail to approach these transformations with ethical clarity and educational foresight, we may find ourselves not masters of technology, but subjects within its design. The question, therefore, is not whether this new world will emerge, it already has, but whether we will remain 'fully human' within it.

2.7 Problems of Adaptation and Integration in a Digitized World

One of the most urgent challenges accompanying the rise of AI and automation lies in the problem of adaptation, the growing inability of workers, industries, and even entire economies to keep pace with technological transformation. While earlier industrial revolutions primarily displaced physical labour, the current wave of intelligent automation increasingly targets cognitive and professional work, including data analysis, diagnostics, design, and decision-support functions that were once assumed to require uniquely human judgment (Autor et al., 2003; Dong & McIntyre, 2014). As we move deeper into this transformation, we observe that the disruption is no longer peripheral but structural, reshaping the very definition of expertise and value in STEM-intensive economies.

For many workers, this shift manifests as displacement, deskilling, or occupational devaluation. AI systems now perform not only routine tasks but also complex pattern recognition and probabilistic reasoning at scales unattainable by humans, challenging long-standing assumptions about human cognitive advantage (Frey & Osborne, 2017). While automation has undoubtedly driven remarkable gains in productivity, these gains have not translated into proportional employment growth. Empirical evidence increasingly suggests that the rate at which technology displaces jobs outpaces the rate at which new, accessible roles are created, particularly for mid-skill and professional workers whose expertise is partially codifiable (Acemoglu & Restrepo, 2020).

This imbalance forces us to confront foundational questions at the heart of the twenty-first-century political economy: Does technological progress ultimately generate more opportunities than it destroys? And if so, for whom? Proponents of technological optimism argue that history demonstrates a long-term pattern of job creation following periods of disruption (Mokyr et al., 2015). Yet this argument increasingly fails to account for the speed, scale, and cognitive reach of contemporary AI systems. Unlike previous technologies, AI does not merely augment labour, it substitutes for it in domains requiring learning, inference, and adaptive reasoning. Without deliberate institutional, educational, and ethical intervention, the promise of future jobs risks becoming inaccessible to those most affected by automation. Until these tensions are addressed with clarity and equity, the human cost of digital transformation will remain one of the defining crises of our age.

At the societal level, the difficulty of adaptation is not merely economic or technological, it is existential. As societies become deeply dependent on digital infrastructures for work, education, healthcare, governance, and social interaction, survival itself increasingly depends on digital literacy, cognitive adaptability, and ethical foresight. We are no longer adapting technology to human life alone; we are adapting human life to technological systems (Floridi, 2014). This shift compels us to reconsider what it means to flourish as human beings in environments shaped by algorithmic mediation and machine intelligence. Without intentional strategies to guide this transition, AI-driven automation risks amplifying social and global inequalities. Access to digital skills, computational literacy, and decision-making power is unevenly distributed, creating a widening gap between a technologically empowered elite and those rendered structurally dependent or excluded (WEF, 2020). From a STEM practitioner's perspective, this raises a critical responsibility: innovation without inclusion is not progress, but stratification.

Embedded within these adaptation challenges are ethical questions of unprecedented scope, particularly concerning how human intellectual agency, responsibility, and authority are redistributed in increasingly networked and automated systems. These challenges can be understood across five interrelated dimensions:

(i) Sharing of digital resources, which raises questions of distributive justice, access to data, computational infrastructure, and opportunities for participation in innovation ecosystems.
(ii) Sharing in private decision-making, where algorithmic systems increasingly mediate corporate strategy, hiring, risk assessment, and productivity, often with limited transparency or accountability.
(iii) Sharing in public decision-making, involving tensions between centralized algorithmic governance and democratic participation, particularly as governments deploy AI in policing, welfare allocation, and public administration.
(iv) Sharing in respect, which speaks to the psychological and social dimensions of adaptation, how individuals retain dignity, purpose, and professional identity when machines outperform them in valued cognitive domains.
(v) Sharing of digital power, encompassing leadership, trust, epistemic authority, and the ethical stewardship of technology in shaping collective futures.

Together, these dimensions constitute the moral architecture of a digitized society, one in which the boundaries between human and machine, private and public, individual and collective are increasingly blurred. For STEM professionals, navigating this terrain requires more than technical competence; it demands ethical literacy and systems thinking. The challenge of adaptation is further compounded by the opacity and complexity of digital systems themselves. Machine learning models, particularly deep neural networks, often operate as black boxes, making their decisions difficult even for experts to fully interpret (Burrell, 2016). On one level, the task before us is human development: cultivating cognitive flexibility, ethical reasoning, and emotional resilience suited to collaboration with intelligent machines.

At another level, it is about human survival in efficiency-driven environments that threaten agency, meaning, and social cohesion. These imperatives, development and survival, are not separable. How we educate STEM students, retrain professionals, govern innovation, and design intelligent systems will determine whether digital transformation enhances human capabilities or accelerates alienation. As we become more deeply woven into the digital fabric, the problems of adaptation and integration will test not only our technical ingenuity, but our moral imagination, the capacity to envision futures in which technology amplifies, rather than erodes, what it means to be human.

2.8 Transforming Tax Billing Policies to Align with Technological Change

In light of the foregoing, there is an urgent and unavoidable need to reform government tax systems to keep pace with the rapid technological transformations reshaping labour markets worldwide. These transformations are not merely about efficiency or innovation; they are fundamentally about power, ethics, and the future of human work. As automation, robotics, and AI increasingly displace both skilled and unskilled workers, we are compelled to ask whether the dominant narrative of technological progress, centred on cost reduction, productivity gains, and shareholder value, adequately accounts for its broader social consequences. *Is the goal of automation truly confined to economic efficiency, or are we witnessing a deeper reconfiguration of the moral and fiscal foundations of work itself?*

There is no question that AI and advanced automation can collaborate with human labour to enhance production, reduce error, and save time. A substantial body of research demonstrates that, under certain conditions, automation complements labour and raises productivity (Aghion et al., 2017). Yet this optimistic framing becomes increasingly fragile when automation systems move beyond routine tasks into non-routine cognitive and decision-making domains. Contemporary AI systems now perform activities, diagnosis, financial assessment, customer interaction, and even creative synthesis, that once conferred professional identity, dignity, and purpose upon human workers (Frey & Osborne, 2017; Nazareno & Schiff, 2021).

To celebrate aggregate economic gains while dismissing the lived reality of job displacement, income erosion, and occupational insecurity is an act of intellectual bad faith. It reflects what critics have described as a techno-utopian bias, one that externalizes social costs while internalizing private gains (Zuboff, 2019). More troubling still, these technologies are increasingly deployed in ways that enable corporations to reduce their tax liabilities, effectively shifting the fiscal burden away from capital and onto labour and consumers. This raises a blunt but unavoidable question: *Who truly benefits from automation, and who pays for its consequences?* Humans, after all, pay taxes on their income, consumption, and labour.

Automated systems, by contrast, generate economic value without directly contributing to the public revenues that sustain the infrastructures, education systems, transport networks, legal frameworks, digital connectivity, upon which they depend. Consider the banking sector, a frequently cited example of successful digitization. Mass job cuts are routinely justified in the name of efficiency, as human tellers are replaced by increasingly sophisticated Automated Teller Machines and AI-driven digital platforms (see Fig. 2.4). These systems now execute complex financial

Fig. 2.4 Automated teller machines as functional components of a bank

operations once performed by skilled workers, yet they do not pay income tax, nor do they contribute directly to social security systems. The question, therefore, is not rhetorical but structural: *Does the ATM pay taxes? And if not, should it—or more precisely, its corporate owner—contribute fiscally in proportion to its economic participation?*

From a public finance perspective, the answer is increasingly difficult to evade. At present, firms enjoy substantial labour cost savings while also benefiting from reduced payroll tax obligations and tax deductions that would otherwise have flowed through employee income taxes. Meanwhile, displaced workers are pushed into lower-paying, more precarious forms of employment, or excluded from the labour market altogether. Empirical evidence suggests that this dynamic contributes to widening inequality and a declining labour share of national income (Acemoglu & Restrepo, 2020; World Bank, 2016). We therefore argue that governments must urgently revisit and recalibrate tax policies to account for the revenue losses induced by automation, robotics, and AI. Proposals such as automation-adjusted corporate taxation, robot taxes, or value-added levies on automated capital, once dismissed as radical, are now entering serious policy discourse (Krafft et al., 2020). Importantly, the ethical justification is not that machines should be treated as moral agents, but that their owners should assume fiscal responsibility for the social displacement their productive systems generate.

Automation forces us to reconsider the moral and economic contract between humans and work. Whenever automation is introduced, two fundamental questions arise: (1) Who performs which tasks in automated environments? and (2) Who benefits from the value produced? A third question inevitably follows: What is the ethical relationship between human contribution and material reward in a technologically mediated economy? These are not merely technical or economic questions; they are existential ones that strike at the core of social cohesion and democratic legitimacy. For us, the central challenge is ensuring that human labour, both skilled and unskilled, remains recognized, valued, and supported in an increasingly automated world. If governments fail to respond with foresight and courage, we risk entrenching a form of digital feudalism: a system in which capital accumulates wealth, machines perform labour, and human workers are left underemployed, fiscally invisible, and socially marginalized. Aligning tax billing policies with technological change is therefore not an act of economic punishment, but one of fiscal justice and societal preservation. In the absence of such reform, technological progress may continue, but at the cost of the very social foundations upon which it depends.

2.9 Unemployment and the Expanding Reach of Automation

When we reflect on Roald Dahl's *Charlie and the Chocolate Factory* (1964, pp. 11–34), one can be surprised by how vividly a children's story anticipates the social costs of technological change. In the novel, young Charlie learns that his father, Mr. Bucket, has lost his job at a toothpaste factory, where his primary task was to manually screw caps onto toothpaste tubes. His dismissal occurs after the factory installs an automated machine capable of performing the same task more cheaply, efficiently, and consistently. As illustrated in Fig. 2.5, the machine replaces human dexterity with mechanical precision, rendering Mr. Bucket economically redundant. Although Dahl's account is fictional, it metaphorically captures what economists now describe as technological unemployment, the displacement of workers due to technological innovation. Unlike earlier waves of mechanisation that primarily affected manual labour, contemporary automation, driven by robotics,

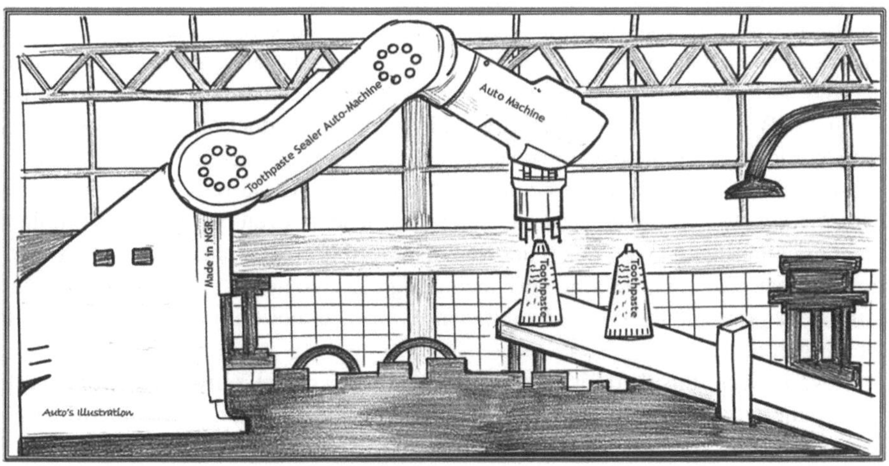

Fig. 2.5 Technological automation replacing human labour

artificial intelligence, and machine learning, extends into cognitive, administrative, and even creative domains. This expansion raises a critical concern. Automation does not merely change how work is performed. It restructures labour markets, alters skill requirements, and risks deepening socio-economic inequality.

Against this literary and economic backdrop, the broader structural implications become clearer. Firstly, technological innovation frequently favours capital over labour. The benefits of automation tend to accrue primarily to those who own productive assets rather than to those who depend on wages for survival. Consequently, productivity gains may increase profits without proportionally expanding employment. The result is not shared prosperity but widening income inequality. Secondly, the speed of adoption intensifies the problem. Automation is now introduced faster than workers can retrain or reposition themselves within evolving labour markets. Unlike previous industrial revolutions, which unfolded over generations and allowed gradual social adaptation, the digital transformation operates within decades. This compressed timeframe reduces the capacity for institutional adjustment and individual reskilling.

In this context, Fig. 2.5 represents more than the replacement of a single worker. It symbolises a structural reordering of economic priorities in which efficiency gains are elevated above employment stability. The figure can therefore be interpreted as reflecting a broader economic logic. Firms adopt automation to reduce labour costs and enhance productivity. In pursuing these objectives, they often externalise the social consequences. These consequences include unemployment, wage stagnation, and worker displacement. Thus, Mr. Bucket's dismissal is not merely a fictional episode. It becomes a philosophical illustration of a persistent tension within capitalist modernity, namely whether technological progress necessarily promotes collective human flourishing or whether its benefits remain unevenly distributed across society.

However, the analysis would be incomplete if we ignore a central counterargument. It is often asserted that automation does not simply destroy work but also creates new occupations. Indeed, technological innovation has generated roles such as AI engineers, data analysts, robotics technicians, and digital marketers, positions that did not previously exist. From this perspective, automation expands economic frontiers. By increasing productivity and lowering production costs, firms may reduce prices, stimulate consumer demand, and thereby generate new employment opportunities elsewhere in the economy. This view finds narrative support even in *Charlie and the Chocolate Factory*, where Mr. Bucket eventually secures employment maintaining the very machines that displaced him. The philosophical implication is clear. We acknowledge that technological change, though disruptive, may ultimately prove regenerative. In this sense, work is not eliminated; instead, it is reconfigured.

Nevertheless, this optimistic account warrants critical scrutiny. Firstly, while new occupations emerge, they are typically skill-intensive and accessible to a limited segment of the labour force. The transition from routine or manual labour to advanced technological roles requires educational credentials, digital literacy, and institutional support that many displaced workers do not possess. Thus, although job creation occurs, it is neither automatic nor evenly distributed. Secondly, the argument that technology complements rather than replaces labour presupposes that workers can realistically reposition themselves within evolving labour markets.

However, Oxford Economics (2019) reports that displacement frequently leads to downward mobility. For every worker successfully retrained into a higher-skilled technological role, many others move into lower-paying, more precarious employment or exit the labour force altogether. In this respect, transformation does not necessarily imply improvement; it may instead signal stratification. Thirdly, the prevailing policy discourse that prioritises reskilling and upskilling as sufficient remedies rests on a narrow economic rationality. It assumes that labour markets are flexible, education systems are responsive, and individuals possess equal adaptive capacity. Philosophically, this reflects a meritocratic ideal. However, social reality is structured by inequality. Access to quality education, geographic mobility, financial resources, and digital infrastructure remains uneven, particularly in developing contexts. As a result, the burden of adaptation falls disproportionately on those least equipped to bear it.

Moreover, the macroeconomic chain effects cannot be ignored. If large segments of the population remain unemployed or underemployed, aggregate purchasing power declines. Reduced income weakens consumer demand, which in turn undermines the very growth dynamics that automation is supposed to stimulate. In economies already marked by structural unemployment, including many African contexts, persistent job displacement may intensify social instability. Where individuals are excluded from formal economic participation and face limited survival options, informal economies and criminal activities can expand.

While automation is not the sole cause of such outcomes, labour market exclusion can exacerbate existing socio-economic vulnerabilities. Therefore, although automation undoubtedly generates new forms of work, the crucial philosophical question concerns distribution rather than mere creation. *Who gains access to these new opportunities, and who remains excluded?* The evidence suggests that without deliberate institutional safeguards, the benefits of automation concentrate among capital owners and highly skilled workers, while the costs are disproportionately borne by lower-skilled and economically marginalised populations (UNECOSOC, 2017; WEF, 2023). In this light, the case of Mr. Bucket is not merely about occupational transformation. It is about structural inequality in the age of intelligent machines.

As STEM educators and researchers, we view this moment as both a warning and a call to action. STEM students now stand at the vanguard of technological innovation, but with that position comes responsibility. It is no longer sufficient to design intelligent systems solely for performance and efficiency; we must also design them with ethical foresight and social awareness. This requires asking difficult but necessary questions: How do we ensure that the benefits of automation do not come at the expense of human dignity? How do we balance technological efficiency with social equity and inclusion? These are not peripheral concerns. They are central to the legitimacy of technological progress itself. Automation's promise is fundamentally double-edged. While it enhances productivity and economic output,

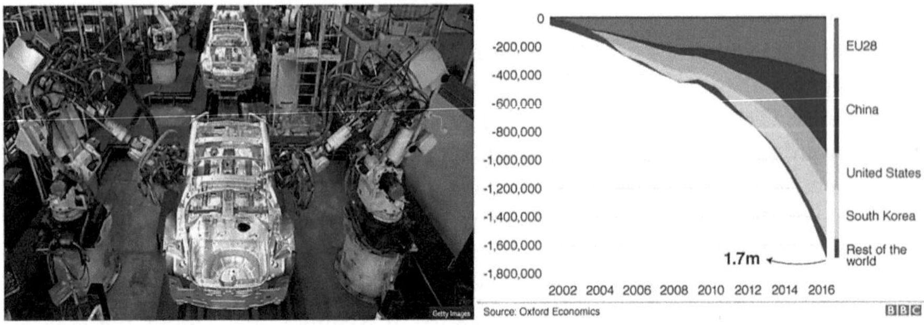

Fig. 2.6 Job losses due to automation. Source information: Retrieve from https://www.bbc.com/news/business-48760799

it simultaneously destabilizes livelihoods, exacerbates income inequality, and widens social divides. An Oxford Economics report projected that up to 20 million manufacturing jobs could be displaced globally by robots by 2050, with particularly severe impacts in regions dependent on routine and semi-skilled labour (see Fig. 2.6). The scale of this disruption challenges any simplistic reliance on reskilling as a universal remedy. It reminds us that the future of work cannot be addressed through technical solutions alone; it demands moral, educational, and institutional responses.

For this reason, we argue that STEM education itself must undergo a profound transformation. Rather than focusing exclusively on technical proficiency, STEM curricula must cultivate reflective practitioners, individuals capable of integrating ethical reasoning, critical thinking, and argumentation into their scientific and engineering practice. The future workforce will not be sustained by efficiency alone. Automation may optimize processes, but it cannot generate meaning. Meaning arises from human deliberation, from our capacity to question assumptions, weigh competing values, and justify decisions in contexts of uncertainty. AI and other digital technologies, no matter how advanced, cannot engage in genuine argumentation. They can process vast datasets, detect patterns, and simulate forms of reasoning, but they cannot experience the moral weight of their conclusions. Argumentation is not merely formal logic; it is an act of judgment grounded in ethos, pathos, and logos, credibility, empathy, and reason. It requires contextual understanding, moral imagination, and self-reflection, qualities that no algorithm can authentically reproduce (Dreyfus, 2009; Floridi, 2014).

For STEM students, then, the future belongs not simply to those who code the fastest, but to those who think the deepest. Argumentation will remain a cornerstone of intellectual survival in an automated world. We must learn to think with technology, not as technology. While AI may assist us in analyzing data, modeling scenarios, or refining claims, it cannot, and must not, replace human critical agency. Automation will continue to advance, but our humanity must advance alongside it. The capacity to reason, to question, and to argue ethically remains both our most powerful form of resistance and our most enduring contribution to the future of intelligent life. Addressing this challenge will be one of the defining tasks for the next generation of innovators, demanding not only technical expertise but a deep understanding of the ethical, economic, and social consequences of the systems they build.

2.10 Argumentation as the Foundation of Critical Thinking, Persuasion, and Ethical Decisions

As we consider the expanding influence of frontier technologies in contemporary society, we are increasingly convinced that the capacity to communicate, reason, and persuade remains one of humanity's most irreplaceable strengths. Research on artificial intelligence and automation consistently demonstrates that while AI systems can process language, identify patterns, and generate syntactically coherent responses, they do so without understanding, intention, or moral accountability (Dong & McIntyre, 2014; Frey & Osborne, 2017). This distinction is critical. Although frontier technologies can imitate debate or rhetorical structure, they lack the human capacity to persuade in the deeper sense, that is, to inspire trust, negotiate shared meaning, and build consensus among people with diverse values and lived experiences.

The automation literature makes clear that contemporary AI systems are optimized for efficiency and prediction rather than interpretation or ethical judgment (Graetz & Michaels, 2018). Argumentation, by contrast, is an interpretive human practice. It integrates logical reasoning with emotional awareness, contextual sensitivity, and moral responsibility. In our teaching and research, we therefore emphasize that genuine argumentation extends far beyond assembling coherent claims. It involves engaging others as moral agents rather than as data points. This difference helps explain why human persuasion continues to outperform computational imitation, even as AI becomes increasingly fluent. Some commentators argue that advances in natural language processing will soon allow AI systems to replace human communicators altogether (Bender et al., 2021; Floridi et al., 2018; Krafft et al., 2020). Indeed, AI already drafts persuasive reports, marketing narratives, and policy analyses. However, we must not equate linguistic fluency with cognitive or ethical agency. Frey and Osborne (2017) emphasize that while automation can substitute for tasks that are routine or rule-based, it struggles with activities requiring judgment under uncertainty, social intelligence, and ethical reasoning.

Persuasion is not merely the mechanical arrangement of words; it arises from human presence, the ability to sense nuance, adapt to emotional context, and assume responsibility for the consequences of one's arguments. This is not a sentimental defence of humanity, but a philosophical and empirical one. Humans persuade because they can care, doubt, and imagine alternative futures. AI systems, by contrast, do not deliberate about the moral implications of their outputs. As Brynjolfsson and McAfee (2014) observe, intelligent machines excel at pattern recognition and scale, but they do not possess conscience or intentionality. Argumentation therefore remains a distinctly human act, one that persists even as AI becomes increasingly capable of linguistic and analytical mimicry.

The ethical stakes of this distinction become especially visible as AI systems are embedded in decision-making processes across governments, corporations, and scientific institutions. Research on automation repeatedly warns that algorithmic systems are designed to optimize efficiency, not justice (Goos et al., 2014; Graetz & Michaels, 2018). We often ask our students a fundamental question that follows directly from this insight: *If AI systems are designed to optimize outcomes, who ensures that those outcomes are fair, humane, and socially legitimate?* Efficiency, as the literature makes clear, is not synonymous with ethics. Argumentation plays a critical role in addressing this gap. Unlike automated optimization, argumentation requires deliberation, justification, and accountability. It forces decision-makers to articulate reasons, confront competing values, and revise positions in response to critique. For STEM students, this makes argumentation a core professional competence rather than a supplementary skill. As automation reshapes professional roles, the capacity to reason ethically and argue responsibly becomes increasingly central to maintaining human agency within technological systems (Frey & Osborne, 2017).

The research surveyed in Artificial Intelligence, Automation and Robotics further demonstrates that many of the ethical dilemmas produced by automation, bias, accountability, transparency, and responsibility, are not reducible to formal logic or computational calculation (e.g., Dong & McIntyre, 2014; Latonero, 2018; Nazareno & Schiff, 2021). They are characterized by uncertainty, ambiguity, and value conflict. In such contexts, the human capacity to reason argumentatively, to deliberate, empathize, and take responsibility, remains indispensable. We therefore remind our students that as future engineers, programmers, and scientists, they are not merely builders of technology. They are its moral architects. Embedding ethical considerations into AI systems is not an optional design feature. It is a professional and civic obligation. Technical solutions alone cannot resolve questions of fairness or legitimacy. These require philosophical reflection and sustained argumentative engagement. Through argumentation, students learn to challenge assumptions, justify decisions, and remain open to ethical revision, capacities that are essential in a world where technological power increasingly outpaces moral consensus.

In the final analysis, the automation literature is clear that while AI can simulate aspects of human cognition, it cannot replace human conscience. Argumentation embodies that conscience. It demands empathy, moral reasoning, and proportional judgment rooted in lived experience rather than computation. For this reason, we see argumentation not only as a skill for navigating the AI-driven future of work, but as a form of intellectual and ethical resistance, a reminder that our capacity to think, question, and care is what ultimately defines our humanity. Before moving to the next chapter, readers, students, educators, and practitioners alike are invited to confront a set of enduring questions: What forms of reasoning are being strengthened through engagement with intelligent technologies? Which forms of judgment are being deferred or surrendered? And who ultimately bears responsibility for decisions made under conditions of automation?

2.11 Chapter Summary

In this chapter, we have positioned contemporary STEM practice within the profound transformations brought about by artificial intelligence, automation, and digitally mediated decision systems. We have shown that STEM work now unfolds within socio-technical environments in which algorithms, data infrastructures, and intelligent tools actively participate in how problems are framed, evaluated, and acted upon. In such contexts, professional competence can no longer be defined solely by technical proficiency or procedural accuracy. It must include the capacity to reason critically, judge ethically, and remain accountable for decisions made in collaboration with intelligent systems. We have argued that while automation can enhance efficiency and extend analytical reach, it does not, and cannot, replace human judgment, moral responsibility, or interpretive reasoning. As routine cognitive labour is increasingly delegated to machines, the distinctly human contribution to STEM work shifts toward capacities such as ethical discernment, creativity, contextual understanding, and reflective decision-making. For both students preparing to enter STEM professions and practitioners already working within AI-augmented environments, this shift demands a deliberate strengthening of cognitive practices that preserve human agency rather than erode it.

Central to this strengthening is argumentation. Throughout the chapter, we have treated argumentation not as a communicative afterthought or a form of debate, but as a foundational mode of reasoning through which evidence is evaluated, assumptions are challenged, and decisions are justified under conditions of uncertainty. Argumentation enables STEM professionals to engage with intelligent systems critically rather than defer to them unreflectively. It supports collaboration across disciplinary and institutional boundaries and provides the means by which ethical considerations are made explicit rather than implicit or ignored. We have also emphasized the educational implications of these developments. If STEM education is to prepare learners for the realities of contemporary professional practice, it must move beyond technical training alone and cultivate habits of inquiry, dialogue, and ethical reflection. Students must learn not only how to use advanced technologies, but how to question them, interpret their outputs, and defend their judgments responsibly. In this sense, argumentation becomes a core pedagogical and professional practice, linking learning, judgment, and accountability.

2.11.1 Looking Ahead to Chap. 3

Having established argumentation as a foundational cognitive and ethical practice for STEM work in an era shaped by artificial intelligence and automation, we now turn to a more precise question: what exactly do we mean by an argument? Before argumentation can function as a professional competency, it must be clearly understood. In Chap. 3, we therefore move from the why of argumentation to the what. We clarify what constitutes a genuine argument and, equally important, what does not, distinguishing reasoned argument from assertion, disagreement, persuasion, and unstructured exchange. For students preparing to enter STEM fields, this distinction is essential for developing disciplined reasoning rather than rhetorical confidence alone. For practitioners, it provides a shared framework for engaging complex technical and ethical questions collaboratively, without reducing disagreement to conflict or authority. Chapter 3 grounds argumentation in evidence-based reasoning and respectful dialogue, demonstrating how well-formed arguments support learning, inquiry, and professional judgment across scientific and mathematical contexts.

As we will see, argumentation is not about winning or convincing others at all costs; it is about understanding, testing ideas, and remaining open to revision. By examining classroom and disciplinary examples, Chap. 3 shows how thoughtful questioning and dialogic engagement cultivate curiosity, rigor, and intellectual humility, qualities that remain indispensable in both STEM education and professional practice. This chapter thus equips readers with the conceptual clarity needed to engage fully with the argumentation practices that anchor the remainder of this handbook.

2.12 Reflection Questions

1. In what ways might reliance on AI systems enhance your professional judgment, and in what ways might it undermine it? How can you distinguish between appropriate delegation and abdication of responsibility?
2. How does argumentation differ from debate or persuasion in STEM contexts? Why is this distinction especially important in AI-mediated decision-making?
3. If an AI-supported decision results in harm, who should be held accountable, and why? How does argumentation help clarify responsibility in such cases?
4. Which cognitive skills are most at risk of erosion as automation increases in your field? What deliberate practices could help preserve and strengthen them?
5. How can STEM professionals ensure that AI systems remain tools for augmentation rather than authorities that dictate decisions?
6. What changes would you recommend to STEM curricula to better prepare students for ethical and cognitive challenges posed by AI and automation?
7. As intelligent systems become more capable, what do you believe will remain uniquely human about your professional role?

Bibliography

Acemoglu, D., & Restrepo, P. (2020). Robots and jobs: Evidence from US labor markets. *Journal of Political Economy, 128*(6), 2188–2244. https://doi.org/10.1086/705716

Aghion, P., Jones, B. F., & Jones, C. I. (2017). *Artificial intelligence and economic growth* (NBER Working Paper No. 23928). National Bureau of Economic Research. https://www.nber.org/papers/w23928

Anderson, J., & Rainie, L. (2018). *Artificial intelligence and the future of humans*. Pew Research Center. https://www.pewresearch.org/internet/2018/12/10/artificial-intelligence-and-the-future-of-humans/

Arntz, M., Gregory, T., & Zierahn, U. (2016). *The risk of automation for jobs in OECD countries: A comparative analysis* (OECD Social, Employment and Migration Working Papers No. 189). OECD Publishing. https://doi.org/10.1787/5jlz9h56dvq7-en

Autor, D. (2015). Why are there still so many jobs? The history and future of workplace automation. *Journal of Economic Perspectives, 29*(3), 3–30. https://doi.org/10.1257/jep.29.3.3

Autor, D. H., Levy, F., & Murnane, R. J. (2003). The skill content of recent technological change: An empirical exploration. *Quarterly Journal of Economics, 118*(4), 1279–1333. https://doi.org/10.1162/003355303322552801

Bender, E. M., Gebru, T., McMillan-Major, A., & Shmitchell, S. (2021). On the dangers of stochastic parrots: Can language models be too big? In Proceedings of the 2021 ACM Conference on Fairness, Accountability, and Transparency (pp. 610–623). https://doi.org/10.1145/3442188.3445922

Bench-Capon, T. J. M., & Dunne, P. E. (2007). Argumentation in artificial intelligence. *Artificial Intelligence, 171*(10–15), 619–641. https://doi.org/10.1016/j.artint.2007.05.001

Berg, A., Buffie, E. F., & Zanna, L.-F. (2018). Should we fear the robot revolution? (The correct answer is yes). *Journal of Monetary Economics, 97*, 117–148. https://doi.org/10.1016/j.jmoneco.2018.05.014

Bowles, S. (2017). *The moral economy: Why good incentives are no substitute for good citizens*. Yale University Press.

Brynjolfsson, E., & McAfee, A. (2014). The second machine age: Work, progress, and prosperity in a time of brilliant technologies. W. W. Norton & Company.

Brynjolfsson, E., Rock, D., & Syverson, C. (2018). Artificial intelligence and the modern productivity paradox: A clash of expectations and statistics. In A. Agrawal, J. Gans, & A. Goldfarb (Eds.), *The economics of artificial intelligence: An agenda* (pp. 23–57). University of Chicago Press.

Burrell, J. (2016). How the machine 'thinks': Understanding opacity in machine learning algorithms. *Big Data & Society, 3*(1), 1–12. https://doi.org/10.1177/2053951715622512

Dahl, R. (1964). *Charlie and the chocolate factory*. George Allen & Unwin.

Dankbaar, B. (1988). New production concepts, management strategies and the quality of work. *Work, Employment and Society, 2*(1), 25–50. https://doi.org/10.1177/0950017088002001003

Davis, C., Hogarth, T., & Gambin, L. (2012). *Sector skills insights: Advanced manufacturing. Evidence report 48*. London, UK, Commission for Employment and Skills.

Demoly, F., Kim, K. Y., & Horváth, I. (2019). Ontological engineering for supporting semantic reasoning in design: Deriving models based on ontologies for supporting engineering design. *Journal of Engineering Design, 30*(10–12), 405–416. https://doi.org/10.1080/09544828.2019.1633626

Dong, X., & McIntyre, S. H. (2014). The second machine age: Work, progress, and prosperity in a time of brilliant technologies. *Quantitative Finance, 14*(11), 1895–1896. https://doi.org/10.1080/14697688.2014.946440

Dung, P. M. (1995). On the acceptability of arguments and its fundamental role in nonmonotonic reasoning, logic programming, and n-person games. *Artificial Intelligence, 77*(2), 321–357.

Dreyfus, H. L. (2009). On the internet (2nd ed.). Routledge.

Floridi, L. (2014). The fourth revolution: How the infosphere is reshaping human reality. Oxford University Press.

Floridi, L., Cowls, J., Beltrametti, M., Chatila, R., Chazerand, P., Dignum, V., Luetge, C., Madelin, R., Pagallo, U., Rossi, F., Schafer, B., Valcke, P., & Vayena, E. (2018). AI4People—An ethical framework for a good AI society: Opportunities, risks, principles, and recommendations. Minds and Machines, 28(4), 689–707. https://doi.org/10.1007/s11023-018-9482-5

Frank, M. R., Autor, D., Bessen, J. E., Brynjolfsson, E., Cebrian, M., Deming, D. J., Feldman, M., Groh, M., Lobo, J., Moro, E., Wang, D., Youn, H., & Rahwan, I. (2019). Toward understanding the impact of artificial intelligence on labor. *Proceedings of the National Academy of Sciences, 116*(14), 6531–6539. https://doi.org/10.1073/pnas.1900949116

Frey, C. B., & Osborne, M. A. (2015). *Technology at work: The future of innovation and employment*. Oxford Martin School & Citi GPS.

Frey, C. B., & Osborne, M. A. (2017). The future of employment: How susceptible are jobs to computerisation? *Technological Forecasting and Social Change, 114*, 254–280. https://doi.org/10.1016/j.techfore.2016.08.019

Goos, M., Manning, A., & Salomons, A. (2014). Explaining job polarization: Routine-biased technological change and offshoring. *American Economic Review, 104*(8), 2509–2526. https://doi.org/10.1257/aer.104.8.2509

Graetz, G., & Michaels, G. (2018). Robots at work. *Review of Economics and Statistics, 100*(5), 753–768. https://doi.org/10.1162/rest_a_00754

Gries, T., & Naudé, W. (2018). The importance of the informal economy in Africa. *The Journal of Development Studies, 54*(6), 1047–1062. https://doi.org/10.1080/00220388.2017.1347622

Greene, D., Hoffmann, A. L., & Stark, L. (2019). Better, nicer, clearer, fairer: A critical assessment of the movement for ethical artificial intelligence and machine learning. In Proceedings of the 52nd Hawaii International Conference on System Sciences (pp. 2122–2131). https://doi.org/10.24251/HICSS.2019.258

Gutek, B. A., & Winter, S. J. (1990). Computer use, control over computers, and job satisfaction. In S. Oskamp & S. Spacapan (Eds.), People's reactions to technology in factories, offices, and aerospace: The Claremont Symposium on applied social psychology (pp. 121–144). Sage Publications.

Halal, W. E. (1993). The information technology revolution: Computer hardware, software, and services into the 21st century. *Technological Forecasting and Social Change, 44*(1), 69–86. https://doi.org/10.1016/0040-1625(93)90007-T

Hästbacka, D., & Kuikka, S. (2011). Bridging UML-profile based models and OWL ontologies in model-driven development. In *Proceedings of the 7th International Conference on Evaluation of Novel Approaches to Software Engineering* (pp. 40–49). ICEIS, Beijing, China.

Iwuanyanwu, P.N. (2020). Nature of problem-solving skills for 21st Century STEM Learners: What teachers need to know. *Journal of STEM Teacher Education*, 55(1), 27 – 40. https://doi.org/10.30707/JSTE55.1/MMDZ8325

Karpouzis, K. (2024). Artificial intelligence in education: Ethical considerations and insights from ancient Greek philosophy. In *Proceedings of the 13th Hellenic Conference on Artificial Intelligence (SETN '24)* (Article 57, pp. 1–7). Association for Computing Machinery. https://doi.org/10.1145/3688671.368877

Korinek, A., & Stiglitz, J. E. (2018). Artificial intelligence and its implications for income distribution and unemployment. In A. Agrawal, J. Gans, & A. Goldfarb (Eds.), *The economics of artificial intelligence: An agenda* (pp. 349–390). University of Chicago Press.

Krafft, P. M., Young, M., Katell, M., Huang, K., & Bugingo, G. (2020). Defining AI in policy versus practice. In *Proceedings of the AAAI/ACM Conference on AI, Ethics, and Society* (pp. 72–78). Association for Computing Machinery. https://doi.org/10.1145/3375627.3375835

Latonero, M. (2018). *Governing artificial intelligence: Upholding human rights and dignity*. Data & Society. https://datasociety.net/pubs/governing_ai.pdf

Leontief, W. (1974). Structure of the world economy: Outline of a simple input-output formulation. *The American Economic Review, 64*(6), 823–834. http://www.jstor.org/stable/1815236

Leontief, W., & Duchin, F. (1986). *The future impact of automation on workers*. Oxford University Press.

Levin, H. M., & Rumberger, R. W. (1987). Educational requirements for new technologies: Visions, possibilities, and current realities. *Educational Policy, 1*(3), 333–354. https://doi.org/10.1177/0895904887001003003

Manjarrés, Á., Fernández-Aller, C., López-Sánchez, M., Rodríguez-Aguilar, J. A., & Sierra Castañer, M. (2021). Artificial intelligence for a fair, just, and equitable world. *IEEE Technology and Society Magazine, 40*(1), 19–24. https://doi.org/10.1109/MTS.2021.3056292
McKinsey Global Institute. (2017). *Jobs lost, jobs gained: Workforce transitions in a time of automation*. McKinsey & Company.
Minda, J. P. (2019). *Cognition: Thinking, reasoning, and problem solving*. SAGE Publications. https://doi.org/10.4135/9781473920262
Mokyr, J., Vickers, C., & Ziebarth, N. L. (2015). The history of technological anxiety and the future of economic growth: Is this time different? *Journal of Economic Perspectives, 29*(3), 31–50. https://doi.org/10.1257/jep.29.3.31
Nazareno, L., & Schiff, D. S. (2021). The impact of automation and artificial intelligence on worker well-being. *Technology in Society, 67*, 101679. https://doi.org/10.1016/j.techsoc.2021.101679
Nguyen, A., Ngo, H. N., Hong, Y., Dang, B., & Nguyen, B. T. (2023). Ethical principles for artificial intelligence in education. *Education and Information Technologies, 28*, 4221–4241. https://doi.org/10.1007/s10639-022-11316-w
OECD. (2019). *Recommendation of the council on artificial intelligence*. OECD Publishing. https://legalinstruments.oecd.org/en/instruments/OECD-LEGAL-0449
Oxford Economics. (2019, June 26). How robots change the world: What automation really means for jobs and productivity. Oxford Economics. https://www.oxfordeconomics.com/resource/howrobots-change-the-world/
Rumberger, R. W. (1984). The impact of dropping out of high school on schooling and labour market outcomes. *Journal of Labor Economics, 2*(2), 118–137. https://doi.org/10.1086/298123
Schwab, K., & Davis, N. (2018). *Shaping the future of the fourth industrial revolution*. World Economic Forum. New York: Crown Business.
Sony, M., & Naik, S. (2020). Industry 4.0 integration with socio-technical systems theory: A systematic review and proposed theoretical model. *Technology in Society, 61*, 101248. https://doi.org/10.1016/j.techsoc.2020.101248
UNECOSOC. (2017). *Joint meeting of ECOSOC and the second committee on "the future of everything—Sustainable development in the age of rapid technological change"*. http://bit.ly/TheFutureofEverything
United Nations Department of Economic and Social Affairs. (2019). World social report 2020: Inequality in a rapidly changing world. United Nations.
Walker, G. H., Stanton, N. A., Salmon, P. M., & Jenkins, D. P. (2008). A review of sociotechnical systems theory: A classic concept for new command and control paradigms. *Theoretical Issues in Ergonomics Science, 9*(6), 479–499. https://doi.org/10.1080/14639220701635470
Wang, W., & Siau, K. (2019). Artificial intelligence, machine learning, automation, robotics, future of work and future of humanity: A review and research agenda. *Journal of Database Management, 30*(1), 61–79. https://doi.org/10.4018/JDM.2019010104
World Bank. (2016). World development report 2016: Digital dividends. World Bank. https://doi.org/10.1596/978-1-4648-0671-1
World Economic Forum (WEF). (2017). *First sustainable development impact summit: Seven ways the fourth industrial revolution can help the planet*. New York, WEF. https://www.weforum.org/events/sustainable-development-impact-summit-2017
World Economic Forum (WEF). (2020). *The future of jobs report 2020*. World Economic Forum.
World Economic Forum (WEF). (2023). *The future of jobs report 2023*. World Economic Forum.
World Economic Forum (WEF). (2024). *The future of jobs report 2024*. World Economic Forum.
Zohar, D. (1995). *The quantum touch: The power of the new science of quantum physics to transform your life*. Quill.
Zohar, D., & Marshall, I. (1990). *The quantum self: Human nature and consciousness defined by the new physics*. Quill.
Zuboff, S. (2019). The age of surveillance capitalism: The fight for a human future at the new frontier of power. PublicAffairs.

Chapter 3
What an Argument Is, and What It Is Not

3.1 Introduction

Before any scientific claim can be justified, defended, or meaningfully understood, it must first enter the space of argument. Argument is not merely a rhetorical exercise or a vehicle for persuasion; it is the medium through which belief is transformed into knowledge, and observation into understanding. It is the disciplined articulation of reasons, evidence, and warrants that renders claims intelligible, accountable, and open to scrutiny. Without this practice, knowledge remains untested, conviction masquerades as certainty, and rational inquiry is reduced to assertion or dispute. In everyday life, the word "argument" circulates casually, often as a synonym for disagreement, conflict, or verbal confrontation, but this common usage obscures its true epistemic function.

Philosophical and educational scholarship reminds us that an argument, properly understood, is not a quarrel; it is a structured act of reasoning oriented toward justification and comprehension rather than domination or emotional release (Toulmin, 1958; Walton, 2012). In this chapter, we invite STEM practitioners and emerging professionals to pause and reconsider their intuitive assumptions about argumentation. We do so because argumentation is not peripheral to STEM practice; it is foundational. Scientific knowledge advances through argument, through the proposing of claims, the marshalling of evidence, the articulation of warrants, and the critical scrutiny of alternatives. When argumentation is misunderstood as mere confrontation, its epistemic and ethical power is diminished.

For STEM educators and those preparing students for professional practice, the implications of this distinction are profound. When learners are not explicitly taught what counts as an argument, they may come to equate confidence with correctness and aggression with justification. Research in STEM education consistently warns against this conflation, emphasizing that argumentation must be cultivated deliberately through instructional design, modeling, and guided practice (Dönmez et al., 2022; Smyrnaiou et al., 2015). We

P. N. Iwuanyanwu, *Empowering STEM Thinkers Through Argumentation*, Contemporary Trends and Issues in Science Education 69,
https://doi.org/10.1007/978-3-032-27906-4_3

therefore advocate for a conception of argumentation that foregrounds coherence, evidence, and ethical purpose. Argumentation is not about silencing others, but about inviting scrutiny. It is not about asserting authority, but about earning credibility. In STEM professions, where decisions often have far-reaching consequences for society, the ability to argue well is inseparable from professional responsibility. As we move deeper into this handbook, we will continue to return to this foundational insight: argumentation is reasoned persuasion, not emotional confrontation. For students, practitioners, and emerging professionals in STEM, mastering this distinction is essential, not only for advancing knowledge and innovation, but for acting responsibly in a world increasingly shaped by the technologies they design, deploy, and defend. When it is understood as structured reasoning, it becomes a central tool for innovation, responsibility, and professional judgment.

3.2 Argument as Structured Reasoning

We begin by drawing a sharp distinction between argumentation and mere discourse. In everyday exchanges as discussed, individuals frequently share opinions, impressions, anecdotes, or isolated pieces of information without any intention of persuading others through reason. Such exchanges may be expressive, informative, or even emotionally charged, but they are not arguments in the logical or philosophical sense. An argument, in its proper sense, is therefore not a quarrel. It is a structured act of reasoning oriented toward justification and persuasion through understanding rather than domination. To argue is to advance a claim, to provide reasons in its support, and to marshal evidence that renders those reasons credible. Equally important, it is to make visible the logical connections that warrant moving from premises to conclusion (Toulmin, 1958). Toulmin's model, widely adopted across STEM education research (e.g., Amgoud & Prade, 2009; Iwuanyanwu, 2020; Kloser et al., 2018; Ogunniyi, 2022), highlights that claims do not stand on their own. They require data (or evidence), warrants that explain why the data support the claim, and, in more sophisticated arguments, backing, qualifiers, and rebuttals. What distinguishes argumentation from mere talk is not volume, passion, or confidence, but structure, coherence, and purpose. Simply trading opinions, sharing isolated facts, or expressing strong feelings, even passionately, does not amount to argumentation unless there is an explicit attempt to justify a position.

Argumentation demands intellectual design. It requires coherence, relevance, and purpose. Research in STEM education reinforces this view. Studies consistently show that high-quality argumentation involves clearly articulated claims, the use of appropriate evidence, and engagement with alternative explanations or counterarguments (e.g., Erduran et al., 2004; Iwuanyanwu & Ogunniyi, 2020; Osborne, 2010). Kuhn (1992) goes further, arguing that mature reasoning itself can be understood as a form of argument, in which individuals learn to coordinate theory and evidence while evaluating competing perspectives. From this standpoint, argumentation is not an optional classroom activity but a core epistemic practice through which understanding deepens and knowledge becomes defensible.

Research in science and mathematics education reinforces this point. Erduran et al. (2004) demonstrate that high-quality argumentation is characterized by explicit claims, the use of relevant evidence, and engagement with counterarguments. Without these elements, discourse may be lively, but it lacks epistemic force. From this perspective, argumentation is less about winning and more about warranting, about showing why a claim ought to be accepted. For STEM students, this distinction is not merely academic. Argumentation is not an optional add-on to STEM learning; it is the very mechanism through which understanding deepens and knowledge advances. This distinction has direct practical consequences. Engineers are routinely required to justify design decisions under conditions of uncertainty. Scientists must defend interpretations of experimental data in the face of competing explanations. Technologists must argue for the safety, reliability, and social acceptability of systems they develop. In each case, unsupported assertions, no matter how confidently delivered, carry little professional weight. Argumentation functions here as a form of intellectual accountability as it makes reasoning visible, examinable, and open to critique.

In an era increasingly shaped by automation, artificial intelligence, and data-intensive technologies, the capacity to argue well is not merely academic, it is existential. Algorithms can calculate, optimize, and predict with extraordinary speed and accuracy, but they cannot justify values, weigh competing ethical considerations, or explain why one course of action ought to be preferred over another. These tasks remain fundamentally human, and they are accomplished through argumentation. Frontier technologies such as artificial intelligence, CRISPR-based genetic engineering, autonomous systems, and the Internet of Things raise questions that cannot be resolved by technical expertise alone. Should an AI system be allowed to make decisions that affect human freedom or dignity? On what grounds should genetic interventions be regulated? How should competing values such as efficiency, equity, privacy, and safety be balanced in technological design? Addressing such questions requires structured ethical argumentation that integrates empirical evidence with normative reasoning (Sadler, 2004; Walton, 2012).

In light of these demands, this chapter argues that STEM professionals who lack the ability to construct and evaluate arguments are professionally incomplete, because the risk they face is not only one of technical error but of ethical failure. Research on socioscientific reasoning shows that individuals who engage explicitly in argumentation are better equipped to navigate uncertainty, consider multiple perspectives, and revise their thinking in light of new evidence (Evagorou & Osborne, 2013; Jiménez-Aleixandre & Erduran, 2008). This capacity is crucial in contexts where decisions must be made under conditions of complexity and value conflict, and where data alone cannot dictate appropriate courses of action. Argumentation thus serves as a safeguard against the illusion that data, by themselves, are sufficient for sound decision-making. In technologically driven systems, data may be abundant, but meaning is not; it is argumentation that gives data both intellectual coherence and moral significance.

3.3 Recognizing Situations with No Argument

Engaging in dissension or speaking loudly to one another does not necessarily imply that an argument is taking place. To clarify what argumentation is *not*, consider the scenario involving Chris and David (see Fig. 3.1). An eyewitness, Jason, initially described the incident as a "heated argument." Their voices were raised, emotions escalated, and bystanders turned to watch as the confrontation intensified. Eventually, the verbal exchange spilled over into physical violence, prompting others to intervene and separate the two men. At first glance, Jason's description seems plausible, as loud and emotional disagreement is commonly labeled an argument. Curious about the nature of this supposed argument, we ask Jason to recount what was actually said. According to his account, Chris repeatedly shouted fragmented utterances such as, "You… yowl… bay… yes… I'll show you!" David, equally agitated, responded with disjointed exclamations of his own: "Hey… you… no way!" Even after they were physically separated, the verbal sparring continued. Jason recalled that Chris kept insisting, "I will show you," while David claimed that he had already proven his point.

At this moment, a critical question arises: What, exactly, was being argued? What claim was Chris advancing? What position was David defending? What evidence was offered, and how did either party attempt to justify their stance? Jason could provide no further detail. Beneath the noise and aggression, there appeared to be no identifiable point of contention, no discernible claim, and no line of reasoning connecting premises to any conclusion. What unfolded was not argumentation but an emotional clash driven by pride, anger, and a struggle for dominance. From the perspective of argumentation theory, this exchange fails to meet even the most basic

Fig. 3.1 The epistemic nature of argumentation misunderstood

Fig. 3.2 No argument to form reasonable opinions

criteria for an argument. There is no claim to evaluate, no counterclaim to assess, and no reasoning to reconstruct. Walton (2012) characterizes such encounters as eristic disputes, verbal battles aimed at defeating an opponent rather than resolving an issue or advancing understanding. These interactions generate heat but no light. They produce conflict rather than knowledge. This realization is instructive. It is not uncommon for eyewitnesses like Jason to report having observed an "argument," only for closer examination to reveal that no argument, in the logical or philosophical sense, ever took place. Without identifiable claims, evidence, or warrants, there is nothing to analyze, critique, or learn from. As illustrated in Fig. 3.2, the absence of claims, evidence, and warrants leaves us with nothing upon which to form reasoned judgment. The account is epistemically thin. There is no foundation for analysis, critique, or learning.

From the perspective of argumentation scholarship, an argument cannot exist without a claim that is open to challenge and justification (Toulmin et al., 1979; Walton, 2012). In Jason's narrative, no such claim is articulated. We are not told what Chris or David asserted, what position either sought to defend, or what evidence was invoked to support those positions. The absence of these elements renders the exchange epistemically opaque. There is nothing to evaluate, no reasoning to reconstruct, and no basis upon which to locate the source of disagreement. This deficiency is not unique to this case. Rather, it reflects a broader phenomenon observable in many real-world conflicts, where disputes remain unresolved or poorly understood precisely because they lack a foundation in reasoned argumentation.

The encounter between Chris and David thus serves as a striking reminder that conflict without argument is not only unproductive but intellectually empty. Without clear claims supported by evidence and connected through reasoning, disagreement devolves into mere confrontation. Argumentation theory distinguishes sharply between reasoned persuasion and eristic exchange, verbal conflict driven by emotion, power, or ego rather than by a search for justification or understanding (Walton, 2012). Jason's account exemplifies the latter. What is reported as a "heated argument" is, upon closer examination, an episode of noise without reason. The absence of key argumentative components in Jason's account also underscores why students, and indeed all members of society, must cultivate argumentation skills. These skills are not solely the domain of STEM professionals, although they are indispensable in scientific and technological contexts. They are equally vital for anyone seeking to engage constructively in public discourse, investigate competing claims, and challenge misinformation or false narratives. Research on scientific reasoning consistently shows that the ability to identify claims, evaluate evidence, and assess reasoning is central to informed citizenship and responsible decision-making (Jiménez-Aleixandre & Erduran, 2008; Kuhn, 1992).

Building on this insight, it becomes evident that one of the most effective ways to develop argumentation skills is by encouraging individuals to engage critically with diverse and sometimes contradictory sources of information. Through such engagement, STEM students, educators, and professionals learn not merely to consume information, but to interrogate it. Structured argumentation requires evaluating the credibility, reliability, and relevance of evidence, as well as considering alternative explanations and counterclaims. In doing so, individuals develop critical thinking, analytical precision, and intellectual humility, qualities essential for navigating complex scientific and technological issues. Figure 3.3 visually consolidates the central claim of this section by showing us, step by step, what distinguishes genuine argumentation from mere discourse or confrontation. As we read the diagram from the top downward, we are invited to move beyond everyday assumptions about arguing and to see argumentation as a disciplined act of reasoning rather than as raised voices or opposing assertions. At the top of the figure, we see the framing idea "Understanding argumentation beyond mere discourse." This immediately aligns with our earlier discussion of the encounter between Chris and David. What Jason reports as an "argument" is exposed, through this lens, as something else entirely. The diagram makes clear that argumentation is not defined by dissension or intensity, but by structure, purpose, and epistemic responsibility.

On the left-hand side, argumentation is presented as a structured presentation in which a claim is advanced, reasons are offered, and evidence is used to support those reasons. Logical coherence is emphasized, showing that premises must lead to conclusions in a way that can be examined and evaluated. This directly highlights what is missing in Jason's account: there are no identifiable claims, no supporting grounds, and no discernible reasoning. Without these elements, as the figure implicitly demonstrates, there is simply no argument to analyse. On the right-hand side, the figure introduces constructive epistemic discourse, reminding us that argumentation is not an isolated act but a mediation process guided by logic, intellectual design, and shared principles. Here, evidence is not merely asserted but used to support claims, leading to what the

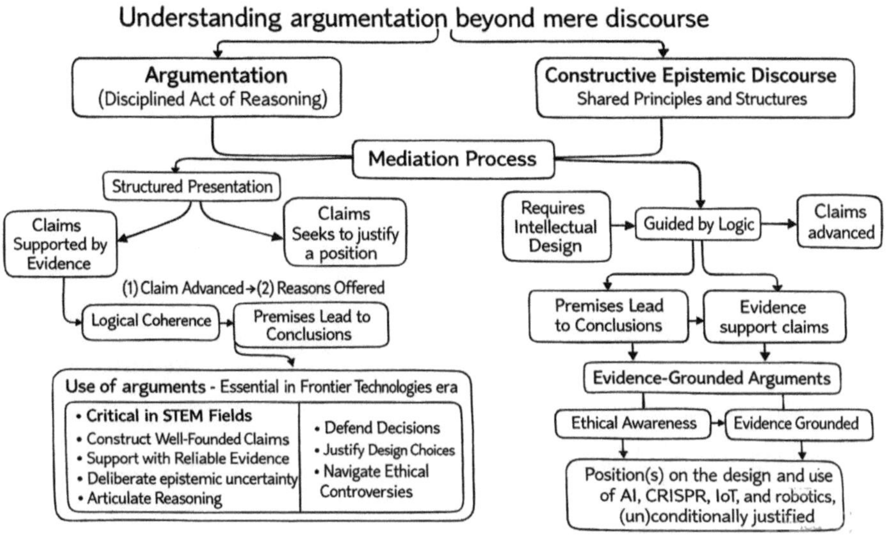

Fig. 3.3 Understanding argumentation as a disciplined act of reasoning

figure identifies as evidence-grounded arguments. This reinforces the idea that argumentation is fundamentally about making reasoning visible and open to scrutiny, rather than asserting dominance or winning a confrontation.

The lower sections of Fig. 3.3 explicitly connect argumentation to contemporary STEM practice. We are shown how argumentation becomes essential in frontier technologies, where professionals must construct well-founded claims, defend decisions, justify design choices, and navigate ethical controversies. The inclusion of ethical awareness and references to AI, CRISPR, IoT, and robotics underscore that argumentation is not only epistemic but moral. Positions on the design and use of such technologies cannot be conditionally or unconditionally justified without evidence-based reasoning. Seen in relation to the Chris and David narrative, Fig. 3.3 helps us recognize situations in which no argument is actually taking place. It clarifies why Jason's account is epistemically thin and why conflicts that lack claims, evidence, and reasoning remain unresolved or misunderstood. At the same time, it points us forward pedagogically. As we encourage students, educators, and professionals to engage critically with diverse and even contradictory sources of information, this figure shows us what structured argumentation looks like in practice. It is through such disciplined reasoning that learners develop critical thinking, analytical skill, and intellectual humility, enabling them to move beyond confrontation toward understanding and informed decision-making.

If we are already familiar with the structure of an argument, Jason's account reveals its deficiencies even more clearly. According to Toulmin's model of argumentation, a reasoned argument comprises several interrelated components: a claim, data (or evidence), a warrant linking the data to the claim, backing to support the warrant, qualifiers indicating the strength of the claim, and potential rebuttals (Toulmin, 1958). Figure 3.4 visually represents this structure. When we compare Figs. 3.1 and 3.2, which depict the confrontation between Chris and David, with

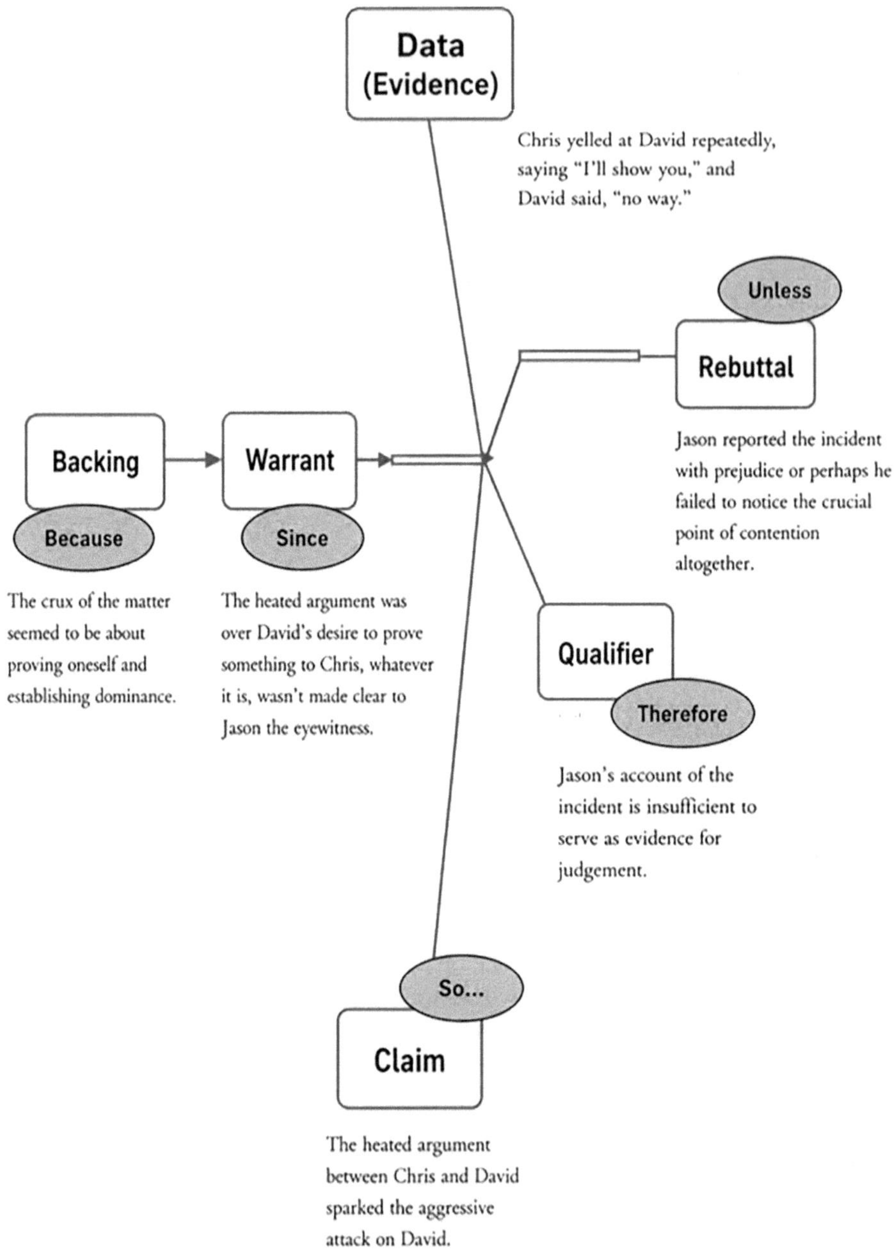

Fig. 3.4 A visual representation of Jason's account within Toulmin's argument structure

Fig. 3.4, the contrast is instructive. Neither Chris nor David demonstrates any of the constituent elements of argumentation, either explicitly or implicitly. There is no identifiable claim, no evidence offered, no reasoning articulated, and no engagement with possible counterpositions. Jason's description therefore highlights the importance of being able to recognize the internal structure of arguments, premises, conclusions, and the reasoning that connects them, if we are to evaluate validity and coherence accurately. Understanding these components enables us to distinguish between arguments that are well-founded and those that are merely rhetorical or performative. This distinction is critical not only in academic contexts but also in professional and public arenas, where the quality of discourse has tangible consequences.

Although Jason refers to the encounter as a "heated argument," his account lacks the organization, coherence, and logical connection that characterize genuine argumentation. The mere presence of fragmented statements, emotional intensity, or verbal exchange does not constitute an argument. Whether argumentation occurs within oneself (intralocutory), between individuals (interlocutory), or among groups (translocutory), each form requires solid grounds and reasons to justify claims and counterclaims. These grounds also determine the appropriate use of modal terms such as probably, possibly, or certainly, which signal the degree of confidence warranted by the evidence. When we map Jason's account onto Toulmin's framework, the absence of logical connections and evidential support becomes unmistakable. The narrative collapses under analytical scrutiny, revealing not an argument, but a misconception about what arguing entails. By contrast, an argument grounded in valid data and supported by sound reasoning is more resilient to criticism, more persuasive to audiences, and more constructive in purpose. When these foundations are missing, as in Jason's account, discourse degenerates into confusion and conflict rather than understanding.

Figure 3.4 helps us visualize this distinction clearly. It shows that when a claim is advanced, it invites questioning: *What reasons support this claim?* In response, data must be provided, and their relevance explained through warrants. These warrants may require backing, may be qualified in scope, and may be challenged through rebuttals. Argumentation thus unfolds as an iterative and dialogic process, one that strengthens understanding by exposing reasoning to critique. In the case of Chris and David, this process never begins. Without claims, data, or warrants, there is no argumentative movement, only confrontation. The shortcomings of Jason's account reflect a broader challenge often observed in eyewitness testimony within legal, scientific, and public contexts. Reports are frequently dismissed not because events did not occur, but because they lack the intellectual structure required to support defensible conclusions. The ability to construct coherent, evidence-driven arguments is therefore not merely an academic exercise; it is central to critical thinking, ethical reasoning, creative problem-solving, and informed judgment.

Analytical, dialectical, and rhetorical forms of argumentation each play a role in this process. Analytical argumentation relies on logical relations between premises and conclusions. Dialectical argumentation emerges through engagement with competing viewpoints. Rhetorical argumentation attends to persuasion and audience. Yet all three depend on the same foundational components: claims, evidence, warrants, and rebuttals. Recognizing these elements allows us to approach complex reasoning with clarity and precision. For this reason, when witnessing a heated exchange, we must resist the impulse to label it an argument too quickly. Instead, we should pause and ask: *Is a claim being made? What evidence supports it? What reasoning connects the evidence to the conclusion? Are counterclaims addressed?* In Jason's case, the absence of answers to these questions confirms that no genuine argument took place. Understanding what an argument is—and what it is not—is therefore essential for distinguishing reasoned discourse from mere verbal confrontation and for participating responsibly in the pursuit of truth, ethical judgment, and meaningful dialogue.

3.4 The Role of Argumentation in STEM Cognition

As emerging technologies continue to transform industries and societies, it is tempting to assume that the primary demand placed on future STEM professionals is ever-greater technical expertise. Certainly, the capacity to design systems, analyse data, and model complex phenomena remains central to STEM practice. Yet, as we pause to consider the nature of the challenges now confronting engineers, scientists, and technologists, it becomes clear that technical proficiency alone is no longer sufficient. Increasingly, STEM professionals are expected to think critically, reason persuasively, and navigate social and ethical dilemmas whose consequences extend far beyond the laboratory or workplace.

Despite this reality, there remains a notable gap in how students are prepared for such demands. Research in science and STEM education consistently shows that when argumentation is not made explicit, students tend to treat scientific knowledge as fixed, authoritative, and beyond question, rather than provisional, contestable, and open to revision (Berland & McNeill, 2010; Osborne, 2010). This tendency encourages passive acceptance rather than reflective engagement. If future STEM professionals are to operate responsibly in complex and uncertain contexts, they cannot approach their work by simply accepting existing claims at face value. Much like Jason's report, which foregrounds the need to question knowledge claims and examine the conditions under which they are produced, students must learn to engage with the world in a reasoned and reflective manner, continually refining their understanding of the situations they encounter.

At this point, it is worth asking why argumentation matters so deeply for STEM learners. For students preparing to enter STEM professions, the ability to form reasoned judgments about complex phenomena and socio-scientific issues that directly affect their communities is not merely an added advantage; it is essential. Research on socio-scientific reasoning shows that argumentation enables learners to integrate scientific evidence with ethical, social, and economic considerations when addressing real-world problems (e.g., Diwu & Ogunniyi, 2012; Evagorou & Osborne, 2013; Sadler et al., 2007). In professional practice, decisions are rarely made on technical grounds alone. STEM students and professionals must therefore acquire genuine argumentation skills that allow them to justify choices, weigh consequences, and explain why one course of action is preferable to another.

This observation leads us to a critical realisation that acquiring technical knowledge, however advanced, is not enough. Students must be able to interrogate the issues before them, challenge underlying assumptions, and engage with multiple perspectives if they are to develop a nuanced understanding of the world they seek to influence. Research on complex and ill-structured problem solving helps clarify why this is so. Problems characteristic of authentic STEM practice cannot be solved through routine procedures or formulaic methods alone; they require judgment, justification, and the comparison of alternative solutions (Adams & Wieman, 2015; Jonassen, 2007). Once we recognise this, an uncomfortable but necessary truth comes into focus: no claim to knowledge, however established, should be accepted uncritically. Jason's eyewitness report captures this point vividly, reminding us that every claim, whether in science, technology, or any other field, must be examined carefully, regardless of how self-evident it may initially appear.

The capacity to ask incisive questions and construct reasoned responses has long been regarded as a defining feature of educated judgment. Philosophical and educational scholarship on argumentation places this capacity at the heart of rational agency and informed decision-making (Siegel, 1995; Toulmin et al., 1979). In earlier eras, such skills were often treated as desirable but secondary to technical mastery. Today, however, they have become indispensable. As automation and artificial intelligence reshape industries, disrupt social systems, and influence personal and professional decisions, the limits of purely technical reasoning become increasingly apparent. While intelligent systems can process data at extraordinary speed and scale, they cannot independently evaluate assumptions, negotiate values, or anticipate ethical consequences. STEM professionals must therefore develop the intellectual courage and analytical rigour to confront uncertainty, scrutinise the tools they use, and reflect critically on the social and ethical implications of their work. In this sense, argumentation is not merely an academic exercise; it is a survival skill in a world where technological decisions have profound human consequences (Longino, 2002; Oreskes, 2019).

The urgency of developing robust argumentative reasoning among future STEM professionals is well supported by empirical evidence. Studies across science, mathematics, and integrated STEM contexts consistently show that argumentation-based instruction leads to deeper conceptual understanding and stronger reasoning than transmission-oriented approaches (e.g., Asterhan & Schwarz, 2007; Erduran & Park, 2023; Walker & Sampson, 2013). What follows from this is an important shift in how we understand learning in STEM. As frontier technologies continue to revolutionise industries, students must be prepared not simply to apply existing knowledge, but to defend their reasoning, explain why particular questions matter, and clarify who may be affected by their decisions. Passive absorption of information is no longer adequate.

The growing complexity of challenges across engineering, technology, and scientific research further reinforces this point. Meaningful engagement with such challenges requires students to move beyond rote learning and to practise sustained critical thinking. This includes evaluating competing viewpoints, constructing well-supported positions, and contributing thoughtfully to problem-solving processes. Research on collective and dialogical argumentation demonstrates that engaging with disagreement and counter-arguments strengthens reasoning quality and improves decision-making rather than undermining it (Chen et al., 2019; Schulz-Hardt et al., 2000). Learning to reason in the presence of disagreement is therefore not a pedagogical risk; it is a professional necessity.

This capacity becomes particularly important in contexts where emerging technologies challenge established norms and assumptions. Developing STEM thinkers involves learning how to weigh competing forms of evidence, critically assess divergent perspectives, and identify the most well-substantiated explanations. Studies of collective argumentation show that such dialogic engagement enhances epistemic vigilance and supports more reasoned judgment (Forman et al., 1998; Krummheuer, 2023). Importantly, this work does not occur in a vacuum. It requires awareness of how social, political, and economic forces shape scientific and technological development.

Artificial intelligence provides a clear illustration of this interplay. Technical capability cannot be separated from ethical and societal questions surrounding its use. Research on argumentation in socio-scientific contexts shows that learners are better prepared to navigate such complexity when they are explicitly taught to justify claims, consider counterarguments, and reflect on consequences (Henderson et al., 2015; Sadler et al., 2007). Cultivating this capacity to assess, revise, and adapt is therefore essential for STEM professionals working in digitally driven and rapidly evolving environments.

From this perspective, STEM education cannot remain confined to the superficial transmission of subject knowledge. Professionals in STEM fields must probe more deeply into the conceptual and ethical foundations of their disciplines. This goes beyond mastering technical content; it requires sustained engagement with critical analysis and argumentation. Research indicates that students who participate in argument-driven inquiry come to understand scientific knowledge as tentative, evidence-based, and open to revision rather than as unquestionable fact (Osborne, 2010; Rapanta et al., 2013). As Jason's case illustrates, claims gain credibility not through authority alone, but through their responsiveness to evidence and critique.

Even as frontier technologies continue to advance, it is human judgment, grounded in reasoning, argumentation, and ethical reflection, that ultimately ensures responsible application of knowledge. Scientific inquiry does not occur in isolation from society, and STEM students must learn to recognise how broader social, political, and economic contexts shape the production and use of knowledge. Relying solely on technical expertise, without attending to these dimensions, is inadequate. Studies grounded in dialogical and culturally responsive models of argumentation show that engaging multiple perspectives strengthens epistemic access and supports more responsible decision-making (Diwu & Ogunniyi, 2012; Ghebru & Ogunniyi, 2017).

Incorporating argumentation into STEM education therefore does more than enhance analytical skill. It empowers students to make informed judgments and to act responsibly in complex and interconnected contexts. Figure 3.5 provides a clear illustration of dialectical engagement in a learning environment, where the instructor engages students in episodes of reasoned argument. Educators must align their teaching strategies with the development of twenty-first-century competencies, including critical thinking, argumentative reasoning, creativity, collaboration, and informed decision-making (Iwuanyanwu, 2020). A growing body of empirical

Fig. 3.5 Fostering genuine argumentation skills

research supports the integration of argumentation-based instruction as a catalyst for meaningful learning. Studies by Erduran and Park (2023), Ghebru and Ogunniyi (2017), Iwuanyanwu (2023), and Osborne (2010) show that argumentation fosters active knowledge construction, careful evaluation of evidence, and reasoned explanation rather than passive reception of information.

Through argumentation, students learn to collaborate and communicate effectively with peers, capacities that are essential in scientific and technological environments characterised by teamwork and interdisciplinary cooperation. Structured discussions and debates teach students to articulate ideas clearly, listen attentively to alternative perspectives, and participate respectfully in intellectual exchange. Research on collective argumentation shows that such practices cultivate open-mindedness, intellectual humility, and epistemic responsibility, qualities that lie at the heart of scientific inquiry and innovation (Krummheuer, 2023; Siegel, 1995). Ultimately, these communicative and argumentative capacities enable students to participate meaningfully in scientific discourse, public deliberation, and interdisciplinary problem-solving. In this way, argumentation moves beyond its role as a classroom activity and becomes a defining feature of professional STEM practice. By integrating argumentation into STEM curricula, we prepare students not merely to understand science and technology, but to think with it, reason through it, and act responsibly within an increasingly complex and interconnected world.

3.5 Arguments as a Matter of Principles

When we engage seriously in argumentation, it quickly becomes apparent that not all arguments are equal. Some withstand scrutiny and deepen understanding, while others collapse under the weight of their own inconsistency. What distinguishes these outcomes is not merely rhetorical skill, but the presence, or absence, of guiding principles. The significance of principles in serious argumentation cannot be overstated, because they establish the standards by which the quality, coherence, and integrity of argumentative discourse are judged. At this point, it is helpful to clarify what is meant by arguing from principle. Principle-based argumentation involves a commitment to standards that remain stable across contexts, even when doing so is inconvenient. When individuals argue from principle, they are not simply trying to prevail in a particular exchange; they are attempting to preserve intellectual and ethical consistency over time. This commitment often gives rise to what is known as the ratchet effect. Once a concession is made within a reasoned dialogue, it becomes the new baseline against which subsequent claims are evaluated. Expectations are raised, not lowered. As a result, principled argumentation demands care, because each move carries consequences for future reasoning.

This explains why principled argumentation can feel demanding, particularly in contexts where expediency is rewarded. Yet the alternative, argumentation driven purely by pragmatism or convenience, comes at a significant cost. Many everyday arguments lack substance and coherence precisely because they are not anchored in enduring principles. Jason's eyewitness report provides a useful illustration here. The absence of clear principles leaves claims unsupported, reasoning fragmented, and conclusions unstable. In such cases, argumentation becomes a performance of opinion rather than an exercise in rational justification. Grounding discourse in principles provides a stabilising framework, allowing participants to maintain coherence even when the argument itself is incomplete or contested. Principles also function as moral and intellectual compasses within argumentative exchanges. They guide individuals in maintaining consistency, honesty, and integrity as discussions unfold. For practitioners and students in STEM fields, this guidance is especially important. Professional credibility depends not only on technical competence, but on the ability to justify decisions in ways that align with shared standards of reasoning and ethical responsibility. Consistency in argumentation signals reliability; it shows that one's commitments are not abandoned when challenged or inconvenient.

This is why the establishment of explicit principles of argumentation is so important in educational and professional contexts. Principles such as respect for evidence, fairness in representing opposing views, openness to revision, and intellectual honesty shape not only what is argued, but how argumentation is conducted. In classrooms, laboratories, design teams, and policy discussions, these principles help ensure that argumentative practices remain aligned with the ethical values and professional standards of the STEM disciplines. When such principles are absent, argumentation risks devolving into competition, persuasion without justification, or strategic manipulation. When they are present, argumentation becomes a disciplined practice of reasoning, capable of sustaining both disagreement and trust.

3.6 Win an Argument or Annihilate an Opponent

When people think about argumentation, they often imagine a contest: one side wins, the other loses. This way of framing argument raises an important question. Is the goal of argumentation to defend one's position successfully, or to defeat one's opponent decisively? At first glance, these may seem like variations of the same aim. Yet a closer examination reveals that both orientations can undermine the very purpose of argumentation. Some individuals take satisfaction in "winning" an argument, while others derive gratification from dismantling or humiliating an opposing position. Although these approaches may appear effective in the short term, they are ultimately counterproductive to learning, understanding, and collective problem-solving. When the focus shifts from inquiry to domination, argumentation becomes a zero-sum exercise. Attacking an opponent may create the appearance of rhetorical strength, but it is an intellectually shallow practice that prioritises performance over substance. In contrast, genuine argumentation seeks understanding, not victory.

At this point, it becomes clear that constructive argumentation requires a different orientation. Rather than framing discussions to dismiss or outmanoeuvre opposing views, we must adopt a dialogical approach, one that encourages participants to feel respected, heard, and taken seriously. This does not mean abandoning critique. On the contrary, it requires a more demanding form of engagement: carefully examining alternative perspectives, identifying both strengths and weaknesses, and working toward greater clarity or resolution. For STEM practitioners, this approach mirrors professional practice, where solutions are refined through critique, iteration, and collaboration rather than through rhetorical dominance.

Criticism that targets difference or imperfection without engaging substance only derails meaningful dialogue. Consider, for example, the tendency to dismiss an argument because it fails to meet an arbitrary expectation. Criticising a chicken for not being a "real bird" because it cannot fly, despite possessing feathers and wings, illustrates the irrationality of such reasoning. In argumentation, similar distortions occur when we reject claims not because they are poorly reasoned, but because they do not conform to our preferred framework. Attacking supposed "hidden motives" behind an argument follows the same pattern. It diverts attention away from evaluating the claim itself and replaces analysis with suspicion.

Paradoxically, such behaviour is often celebrated as cleverness in competitive discourse. Yet this form of cleverness reflects superficial engagement rather than intellectual depth. If argumentation is to fulfil its educative and civic purposes, it must move beyond a win–lose mentality toward one of mutual benefit and shared inquiry. This shift requires active listening, empathy, and a willingness to acknowledge valid points even when they originate from opposing positions. These practices cultivate intellectual humility and strengthen the collective pursuit of knowledge. The misuse of fallacious reasoning provides a clear illustration of how destructive argumentation can become. The straw man fallacy, depicted in Fig. 3.6, occurs when an opponent's position is distorted to make it easier to attack. Rather than engaging the strongest version of the argument, the critic constructs a

Fig. 3.6 Straw man fallacy in argument (destructive argument)

weakened caricature. This tactic undermines the integrity of discourse and borders on verbal bullying, as it shifts attention from ideas to misrepresentation. The long-term consequence is the erosion of trust, respect, and openness, conditions essential for any meaningful exchange of ideas.

Instead of embracing distortion and confrontation, argumentation should foster reflection and understanding. By focusing on what is actually being claimed, rather than what is convenient to refute, participants create space for learning and insight. In many cases, apparent disagreements arise not from genuine opposition, but from misunderstanding. What initially appears to be a clash of ideas often turns out to be a semantic misalignment or a partial articulation of a broader belief system. When arguments consist of fragments rather than fully justified claims, discourse degenerates into assertion without evidence, a display of conviction rather than a pursuit of truth. There is also a deeper paradox worth considering. Individuals who adhere rigidly to a particular ideology may, in their efforts to defend it, end up violating the very principles they claim to uphold. Unquestioned allegiance to any doctrine, whether scientific, religious, or political, can produce epistemic blindness, obscuring inconsistencies and suppressing self-critique. In such cases, dialogue transforms into a battleground of identities, where arguments are judged not by their merit but by their alignment with a fixed worldview. The danger lies not in ideology itself, which is an unavoidable feature of human cognition, but in its sanctification beyond question.

Effective argumentation therefore requires the willingness to subject one's own premises to the same scrutiny applied to others. Yet this scrutiny must never be confused with personal attack. Ideas must be challenged; people must be respected. When critique of thought is conflated with critique of personhood, argumentation collapses into polemic. This is why epistemic empathy, the ability to enter another's framework of reasoning without necessarily endorsing it, is so important. Pausing to ask what the real point of contention is often reveals that no substantive conflict exists at all, only differences in emphasis, perspective, or language. Seen in this light, the task of argumentation is not to defeat, but to clarify; not to dominate, but to illuminate. For STEM practitioners and students alike, this orientation transforms argumentation from a contest of words into a collaborative practice of reasoning. It allows disagreement to become productive rather than destructive, and inquiry to replace antagonism. When argumentation is approached in this way, it fulfils its highest purpose: advancing understanding while preserving the dignity of all who participate in the exchange.

3.7 Developing and Practicing Constructive Argumentative Discourse

At this stage of our inquiry, it is important to address a common misconception that the ability to argue constructively is a natural talent possessed by only a few. In practice, constructive argumentative discourse is not a gift; it is a cultivated intellectual virtue. Research in argumentation and science education consistently shows that high-quality argumentation emerges through deliberate development, guided practice, and explicit norms of engagement rather than spontaneous debate (Erduran et al., 2004). For STEM students and professionals alike, learning to argue well is therefore a matter of disciplined formation rather than innate ability. To argue constructively is to treat discourse as a collaborative exploration of truth, not as a contest for dominance. Argumentation scholars emphasise that productive discourse is characterised by shared responsibility for reasoning rather than individual victory (e.g., Asterhan & Schwarz, 2007; Mercier & Sperber, 2011; Ogunniyi & Iwuanyanwu, 2024). From this perspective, developing constructive argumentation involves at least three interrelated commitments: open-mindedness, evidence-based reasoning, and communicative integrity.

We begin with open-mindedness, because without it, argumentation collapses into defence rather than inquiry. Open-mindedness does not require the abandonment of conviction. Rather, it demands the courage to allow one's convictions to be tested through reasoned engagement with alternatives. Philosophers of education have long argued that intellectual growth begins at the point where certainty is unsettled and examined (Siegel, 1995). A mind that cannot accommodate alternative viewpoints may appear principled, but in practice it becomes epistemically closed. Research on dialogical argumentation confirms that learners who are encouraged to listen with the intent to understand, rather than to rebut, demonstrate greater conceptual change and deeper reasoning (Krummheuer, 2023). The second commitment concerns evidence-based reasoning. Constructive argumentation requires that claims be grounded in evidence and linked through coherent reasoning. In STEM contexts, this principle is particularly critical. Argumentation research shows that when students are taught to justify claims using data, warrants, and backing, their understanding of scientific concepts becomes more robust and transferable (Erduran & Jiménez-Aleixandre, 2008). In an era marked by misinformation and ideological polarisation, the disciplined pursuit of evidence is not merely an academic requirement; it is an ethical obligation. It compels us to distinguish between what we believe to be true and what can be justified within a community of reason.

From my own experience teaching STEM students, arguments grounded in data, analysis, and logical coherence not only sharpen cognitive skills but also cultivate epistemic responsibility, the capacity to account for one's beliefs publicly and transparently. This aligns with research showing that argumentation supports students in developing norms of accountability and justification that mirror professional scientific practice (Berland & Reiser, 2009). The third commitment, communicative integrity, is often overlooked but equally essential. Argumentation is not only about what is said, but about how it is said. Studies of classroom and professional discourse demonstrate that respectful communication, clarity of expression, and attentive listening significantly affect the quality of reasoning that emerges (Asterhan & Schwarz, 2007; Osborne, 2010). Active listening, careful articulation, and sensitivity to language reduce misunderstanding and foster mutual respect. Without these communicative conditions, even well-supported arguments can fail to achieve constructive engagement.

From an educational perspective, cultivating these practices transforms the classroom into a microcosm of democratic and scientific life, a space where ideas, rather than individuals, are contested. Through repeated engagement in structured argumentation, students learn to separate ego from evidence and conviction from justification. Some critics argue that emphasising civility risks diluting the passion necessary for transformative argumentation. Indeed, research acknowledges that a degree of tension and discomfort can be pedagogically productive (Levinson, 2006). However, evidence also shows that intellectual rigour and respect are not opposites. On the contrary, high-quality argumentation flourishes where critical intensity is balanced by moral restraint (Erduran & Park, 2023; Siegel, 1995).

Constructive argumentative discourse is therefore both an individual and a collective achievement. Individually, it nurtures critical and creative thinking. Collectively, it sustains democratic dialogue and scientific inquiry. In STEM education, students who engage in principled, evidence-based argumentation are better equipped to navigate the ethical and epistemic challenges posed by emerging technologies. They learn not only to solve problems, but to question assumptions, a skill that is indispensable in an AI-driven world where the boundaries of truth and interpretation are continually renegotiated. Practised in this way, constructive argumentation becomes a form of intellectual citizenship: the capacity to participate meaningfully in the life of reason with both courage and compassion.

3.8 Achieving Constructive Outcome in Argumentation

If constructive argumentation is a cultivated practice, then constructive outcomes are its cumulative result. Such outcomes do not arise by accident, nor do they emerge instantaneously. Research on dialogical reasoning suggests that productive argumentative exchanges develop through systematic and progressive engagement, particularly when participants commit to shared principles and a coherent structure of interaction (Mercier, 2011). Even when interlocutors begin from opposing positions, argumentation can evolve into a meaningful exchange of understanding rather than a contest of wills. At this point, it is important to clarify what we mean by a constructive outcome. It is unrealistic to expect that one party will immediately adopt the other's position or propose a solution that satisfies all involved. Argumentation is inherently iterative. Progress unfolds through incremental shifts in understanding rather than through conversion or surrender. Studies of productive disagreement show that participants are more likely to remain engaged when each stage of the dialogue offers some epistemic gain greater clarity, recognition of complexity, or refinement of one's own position (Asterhan & Schwarz, 2007; Schulz-Hardt et al., 2000).

A brief classroom example helps to illustrate this point. One morning, I encountered two students, Sam and Steve, engaged in a spirited discussion about the English Premier League. Sam, a committed Manchester City supporter, asserted with confidence that his team would inevitably win the league. He supported his claim by citing recent performance statistics, consistency, and expert commentary. His reasoning followed a familiar empirical logic that past dominance serves as a predictor of future success. Steve, an Arsenal supporter, did not simply reject Sam's conclusion. Instead, he introduced a critical counterpoint. He questioned whether historical data alone was sufficient justification for certainty. Factors such as morale, injuries, tactical shifts, and human unpredictability, he argued, complicate straightforward extrapolation. Steve's intervention redirected the exchange from data accumulation to epistemic uncertainty. This move reflects what argumentation scholars describe as productive counter-argumentation, challenging not the conclusion itself, but the strength and scope of the warrant connecting evidence to claim (Toulmin, 2003).

What made this exchange constructive was not that Sam and Steve reached agreement. They did not. Rather, they left the discussion with a clearer understanding of why they disagreed. Research on argumentation emphasises that such outcomes, greater awareness of assumptions, limits of evidence, and alternative interpretations, are central indicators of productive discourse (Osborne, 2010; Rapanta et al., 2013). Argumentation, in this sense, need not culminate in consensus to be valuable. Intellectual transformation, not persuasion, is the relevant measure of success. This raises a deeper philosophical question: *must all arguments aim at consensus?* Argumentation theorists caution that an uncritical pursuit of consensus can suppress valuable dissent and prematurely close inquiry (Longino, 2002). Disagreement, when grounded in mutual respect and critical reasoning, is often epistemically productive. It exposes the limits of our understanding and invites deeper reflection. For STEM education, this insight is particularly important. When instruction is confined to transmitting established knowledge, students may succeed in recall but struggle with reasoning. When learning is framed as an argumentative dialogue, students learn to question authority, evaluate claims, and justify conclusions, capacities essential for scientific progress.

3.9 Chapter Summary

In this chapter, we have explored a foundational idea that understanding what an argument is, and what it is not, is essential for meaningful discourse. As we have seen, many of the tensions that arise in daily communication, scientific inquiry, and education stem from the failure to distinguish arguments from mere assertions or quarrels. By recognising an argument as a reasoned attempt to justify a claim, we equip ourselves to evaluate not only its validity and coherence but also the ethical intentions behind it (Toulmin, 2003). This distinction is more than technical; it is the lens through which we can navigate and make sense of complex claims in STEM and beyond. We have also argued that the power of argumentation is most fully realised when it is guided by principles rather than by the desire to "win." Principles such as fairness, rationality, consistency, and respect serve as standards for evaluating the strength of our reasoning. When we commit to these principles, we cultivate intellectual virtues such as honesty, openness, and moral discernment. In contrast, an adversarial approach, focused on personal victory, risks reducing argumentation to a contest of rhetorical skill, eroding both its educative and ethical value.

As we turn our attention to STEM education, it becomes clear that argumentation has transformative potential. When students engage in argument, testing hypotheses, evaluating evidence, and constructing explanations, they are doing more than memorising facts; they are participating in the authentic practice of science. We can see, then, that argumentation is not an optional enrichment but a bridge between knowledge and understanding, allowing students to explore how scientific ideas are developed, contested, and refined. In other words, by arguing, we are thinking like scientists. At this point, we might anticipate a concern: *does emphasising argumentation, with its inherent uncertainty and dissent, risk undermining confidence or slowing instruction?* We acknowledge this, yet research suggests that moments of intellectual discomfort are precisely where understanding deepens (Posner et al., 1982; Nussbaum & Edwards, 2011). When we encounter doubt and tension, we are invited to reconcile competing ideas and refine our reasoning. In this way, intellectual discomfort is not a barrier, it is the catalyst for growth.

Finally, we must remember that argumentation is more than a pedagogical strategy; it is an ethical stance toward knowledge and others. To argue constructively is to honour the dignity of reason, to recognise the legitimacy of difference, and to remain open to shared understanding. When we guide STEM students to argue critically, respectfully, and persuasively, we are also teaching them to participate responsibly in a pluralistic, technologically complex world. By embracing argumentation as both method and ethos, we move closer to an educational vision where inquiry replaces indoctrination, understanding replaces domination, and truth becomes a collective pursuit rather than a private possession. If we take this vision seriously, we do more than prepare students to know STEM knowledge and technical skills. We prepare them to think, reason, and act as thoughtful, responsible professionals in an ever-changing world.

3.9.1 Looking Ahead to Chap. 4

In this chapter, we have taken care to slow down and examine what we often take for granted. By disentangling argumentation from quarrel, confrontation, and rhetorical performance, we have clarified that an argument is a disciplined act of reasoning grounded in claims, evidence, warrants, and guiding principles. We have seen that without these elements, disagreement collapses into noise rather than knowledge, and persuasion loses its epistemic and ethical force. This clarification is not merely conceptual; it is foundational. Before we can ask students and emerging professionals to argue well in STEM contexts, we must first be clear about what it means to argue at all. Having established this groundwork, we are now ready to move forward.

In the next chapter, we turn our attention to the epistemic practices that make argumentation possible and meaningful. We examine reasoning, intellect, and evidence not as isolated skills, but as mutually reinforcing capacities that guide responsible judgment in STEM education and practice. We explore how deductive and inductive reasoning work together, forming a cognitive movement between logic and understanding that allows individuals to navigate complexity without retreating into either rigid certainty or uncritical relativism. As we proceed, we also deepen the ethical dimension of argumentation that has surfaced throughout this chapter. Through concrete illustrations, including a vignette of a student who insists on evidence despite social pressure, we confront the reality that argumentation is never value-neutral. To argue well is to exercise intellectual courage, humility, perseverance, and reflective judgment. In this sense, argumentation emerges not only as a method of justification, but as a way of being in relation to knowledge and to others.

If this chapter has helped us recognise what an argument *is* and what it *is not*, the question that naturally follows is: *how do individuals actually develop, sustain, and evaluate reasoned arguments in situations of uncertainty, pressure, and competing claims?* This is the central concern of Chap. 4. Chapter 4 therefore invites us to see argumentation as more than a technical competence. It asks us to consider it as a pedagogical ethos, one that connects reasoning to decision-making, scientific inquiry to ethical deliberation, and evidence to responsibility. As we move forward, we do so with the understanding developed in this chapter that structured argumentation is not an obstacle to learning, but the very means through which understanding, judgment, and professional integrity are formed. Together, this chapter and Chap. 4 advance a shared inquiry into how STEM education can cultivate individuals capable of reasoning carefully, deciding responsibly, and participating thoughtfully in scientific and societal problem-solving.

3.10 Reflection Questions

1. In everyday language, the word *argument* is often associated with conflict or confrontation. How has this chapter challenged your prior understanding of what an argument is? In what ways might this shift influence how you engage in discussions in academic or professional contexts?
2. Consider a recent disagreement you observed or participated in. Using Toulmin's model, were there identifiable claims, evidence, and warrants present? If not, what was missing, and how did that absence affect the outcome of the exchange?
3. The chapter argues that argumentation is foundational to STEM practice rather than an optional add-on. Why is structured argumentation particularly important in contexts involving emerging technologies such as AI, biotechnology, or autonomous systems?
4. Reflect on the distinction between *winning an argument* and *arguing constructively*. How might a desire to win undermine understanding, collaboration, or ethical responsibility in STEM-related decision-making?
5. Jason's eyewitness account illustrates a situation where no genuine argument took place. Why is the ability to recognise the *absence* of argumentation as important as the ability to evaluate strong arguments?
6. What role do principles such as consistency, fairness, and respect for evidence play in sustaining meaningful argumentative discourse? How might abandoning these principles affect professional credibility in STEM fields?
7. The chapter emphasises that argumentation involves intellectual accountability. What does it mean to make one's reasoning visible and open to critique, and why might this be uncomfortable yet necessary?
8. How does the development of constructive argumentative discourse contribute to intellectual humility and ethical judgment? Can you identify situations where these virtues would be critical in STEM education or practice?
9. In what ways does argumentation serve as a safeguard against the assumption that data alone are sufficient for sound decision-making?
10. After engaging with this chapter, how would you explain to a student or colleague the difference between an argument, a debate, and a quarrel?

References

Adams, W. K., & Wieman, C. E. (2015). Analyzing the many skills involved in solving complex physics problems. *American Journal of Physics, 83*(5), 459–467. https://doi.org/10.1119/1.4913923

Amgoud, L., & Prade, H. (2009). Using arguments for making and explaining decisions. *Artificial Intelligence, 173*(3–4), 413–436. https://doi.org/10.1016/j.artint.2008.11.006

Asterhan, C. S. C., & Schwarz, B. B. (2007). The effects of monological and dialogical argumentation on concept learning in evolutionary theory. *Journal of Educational Psychology, 99*(3), 626–639. https://doi.org/10.1037/0022-0663.99.3.626

Berland, L. K., & Reiser, B. J. (2009). Making sense of argumentation and explanation. *Science Education, 93*(1), 26–55.

Berland, L. K., & McNeill, K. L. (2010). A learning progression for scientific argumentation: Understanding student work and designing supportive instructional contexts. *Science Education, 94*(5), 765–793. https://doi.org/10.1002/sce.20402

Chen, Y.-C., Benus, M. J., & Hernandez, J. (2019). Managing uncertainty in scientific argumentation. *Science Education, 103*(5), 1235–1276. https://doi.org/10.1002/sce.21527

Diwu, C. T., & Ogunniyi, M. B. (2012). Dialogical argumentation instruction as a catalytic agent for the integration of school science with Indigenous Knowledge Systems. *African Journal of Research in Mathematics, Science and Technology Education, 16*(3), 333–347. https://doi.org/10.1080/10288457.2012.10740749

Dönmez, İ., Gülen, S., & Ayaz, M. (2022). Impact of argumentation-based STEM Activities on ongoing STEM motivation. *Journal for STEM Education Research, 5*, 78–101. https://doi.org/10.1007/s41979-021-00062-2

Erduran, S., & Jiménez-Aleixandre, M. P. (Eds.). (2008). Argumentation in science education: Perspectives from classroom-based research. Springer. https://doi.org/10.1007/978-1-4020-6670-2

Erduran, S., & Park, W. (2023). Argumentation in physics education research: Recent trends and key themes. In M. F. Taşar & P. R. L. Heron (Eds.), *The international handbook of physics education research: Learning physics* (pp. 16–32). AIP Publishing.

Erduran, S., Simon, S., & Osborne, J. (2004). TAPping into argumentation. *Science Education, 88*(6), 915–933. https://doi.org/10.1002/sce.20012

Evagorou, M., & Osborne, J. (2013). Exploring young students' collaborative argumentation within a socioscientific issue. *Journal of Research in Science Teaching, 50*(2), 209–237. https://doi.org/10.1002/tea.21076

Forman, E. A., Larreamendy-Joerns, J., Stein, M. K., & Brown, C. A. (1998). "You're going to want to find out which and prove it": Collective argumentation in mathematics classrooms. *Learning and Instruction, 8*(6), 527–548. https://doi.org/10.1016/S0959-4752(98)00033-4

Ghebru, S., & Ogunniyi, M. (2017). Pre-service science teachers' understanding of argumentation. *African Journal of Research in Mathematics, Science and Technology Education, 21*(1), 49–60. https://doi.org/10.1080/18117295.2016.1254493

Henderson, J. B., McNeill, K. L., Gonzalez-Howard, M., Close, K., & Evans, M. (2015). Key challenges and future directions for educational research on scientific argumentation. *Journal of Research in Science Teaching, 55*(1), 5–8. https://doi.org/10.1002/tea.21412

Iwuanyanwu, P. N. (2020). Nature of problem-solving skills for 21st century STEM learners: What teachers need to know. *Journal of STEM Teacher Education, 55*(1), 27–40. https://doi.org/10.30707/JSTE55.1/MMDZ8325

Iwuanyanwu, P. N. (2023). When Science Is Taught This Way, Students Become Critical Friends: Setting the Stage for Student Teachers. *Research in Science Education, 53*, 1063–1079. https://doi.org/10.1007/s11165-023-10122-9

Iwuanyanwu, P. N., & Ogunniyi, M. B. (2020). Effects of dialogical argumentation instructional model on pre-service teachers' ability to solve conceptual mathematical problems in physics. *African Journal of Research in Mathematics, Science and Technology Education, 24*(1), 121–141. https://doi.org/10.1080/18117295.2020.1748325

Jiménez-Aleixandre, M. P., & Erduran, S. (2008). Argumentation in science education: An overview. In S. Erduran & M. P. Jiménez-Aleixandre (Eds.), *Argumentation in science education: Perspectives from classroom-based research* (pp. 3–27). Springer. https://doi.org/10.1007/978-1-4020-6670-2_1

Jonassen, D. H. (2007). *Learning to solve problems: A handbook for designing problem-solving learning environments*. Routledge.

Kloser, M., Wilsey, M., Hopkins, D. W., Dallavis, J. W., Lavin, E., & Comuniello, M. (2018). Dual identities: Organizational negotiation in STEM-focused Catholic schools. *Cultural Studies of Science Education, 13*(2), 549–579. https://doi.org/10.1007/s11422-017-9819-z

Krummheuer, G. (2023). Negotiation, argumentation and participation: Three basic concepts referring to everyday procedures in teaching and learning situations in mathematics classes. *Theory Into Practice, 62*(1), 70–78. https://doi.org/10.1080/00405841.2022.2135901

Kuhn, D. (1992). Thinking as argument. *Harvard Educational Review, 62*(2), 155–178.

Levinson, R. (2006). Towards a theoretical framework for teaching controversial socio-scientific issues. *International Journal of Science Education, 28*(10), 1201–1224. https://doi.org/10.1080/09500690600560753
Longino, H. E. (2002). *The fate of knowledge*. Princeton University Press.
Mercier, H. (2011). Reasoning serves argumentation in children. *Cognitive Development, 26*(3), 177–191. https://doi.org/10.1016/j.cogdev.2010.12.001
Mercier, H., & Sperber, D. (2011). Why do humans reason? Arguments for an argumentative theory. *Behavioral and Brain Sciences, 34*(2), 57–74. https://doi.org/10.1017/S0140525X10000968
Nussbaum, E. M., & Edwards, O. V. (2011). Critical questions and argument stratagems: A framework for enhancing and analyzing students' reasoning practices. *Journal of the Learning Sciences, 20*(3), 443–488. https://doi.org/10.1080/10508406.2011.564567
Ogunniyi, M. B. (2022). Implementing a socioculturally relevant science curriculum: The South African experience. In M. M. Atwater (Eds.), *International handbook of research on multicultural science education*
Ogunniyi, M., & Iwuanyanwu, P. N. (2024). Analysis of Teachers' Perspectives Towards the Use of IKS to Improve STEM Education for Sustainable Development. *African Journal of Research in Mathematics, Science and Technology Education, 28*(3), 319–329. https://doi.org/10.1080/18117295.2024.2352980
Oreskes, N. (2019). *Why trust science?* Princeton University Press. https://doi.org/10.2307/j.ctvfjczmg
Osborne, J. (2010). Arguing to learn in science: The role of collaborative, critical discourse. *Science, 328*, 463–466. https://doi.org/10.1126/science.1183944
Posner, G. J., Strike, K. A., Hewson, P. W., & Gertzog, W. A. (1982). Accommodation of a scientific conception: Toward a theory of conceptual change. *Science Education, 66*(2), 211–227. https://doi.org/10.1002/sce.3730660207
Rapanta, C., Garcia-Mila, M., & Gilabert, S. (2013). What is meant by argumentative competence? An integrative review of methods of analysis and assessment in education. *Review of Educational Research, 83*(4), 483–520. https://doi.org/10.3102/0034654313487606
Sadler, T. D. (2004). Informal reasoning regarding socioscientific issues: A critical review of research. *Journal of Research in Science Teaching, 41*(5), 513–536. https://doi.org/10.1002/tea.20009
Sadler, T. D., Barab, S. A., & Scott, B. (2007). What do students gain by engaging in socio-scientific inquiry? *Research in Science Education, 37*, 371–391. https://doi.org/10.1007/s11165-006-9030-9
Schulz-Hardt, S., Frey, D., Lüthgens, C., & Moscovici, S. (2000). Biased information search in group decision making. *Journal of Personality and Social Psychology, 78*(4), 655–669. https://doi.org/10.1037/0022-3514.78.4.655
Siegel, H. (1995). Why should educators care about argumentation? *Informal Logic, 17*(2), 159–176. https://doi.org/10.22329/il.v17i2.2424
Smyrnaiou, Z., Petropoulou, E., & Sotiriou, M. (2015). Applying argumentation approach in STEM education: A case study of the European student parliaments project in Greece. *American Journal of Educational Research, 3*(12), 1618–1628.
Toulmin, S. (1958/2003). *The uses of argument*. Cambridge University Press.
Toulmin, S., Rieke, R. D., & Janik, A. (1979). *An introduction to reasoning* (3rd ed.). Macmillan.
Walker, J. P., & Sampson, V. (2013). Learning to argue and arguing to learn: Argument-driven inquiry as a way to help undergraduate chemistry students learn how to construct arguments and engage in argumentation during a laboratory course. *Journal of Research in Science Teaching, 50*(5), 561–596. https://doi.org/10.1002/tea.21082
Walton, D. (2012). *Fundamentals of critical argumentation. Cambridge University Press.* https://doi.org/10.1017/CBO9780511807039

Chapter 4
The Epistemic Practices of Developing Reasoned Arguments

4.1 Introduction

In this chapter, we turn our attention to the epistemic practices that underpin the development of reasoned arguments in STEM fields. These practices are not peripheral skills acquired after technical mastery; they are foundational to how knowledge is generated, tested, and validated. As scientists, engineers, technologists, and mathematicians, we rely on argumentation not merely to justify conclusions or defend results, but to extend understanding and challenge the limits of what is currently known. Argumentation, in this sense, is not an accessory to scientific reasoning; it is its lifeblood. To appreciate this fully, we must look beyond argumentation as a communicative technique and recognise it as an epistemic activity. Developing a reasoned argument requires more than technical competence. It calls for an intellectual disposition, a willingness to gather evidence carefully, interpret it thoughtfully, and evaluate claims through reflective and dialogic inquiry.

When we examine how arguments function in STEM, we begin to see that reasoning and intellect operate together as twin pillars. Reasoning enables us to analyse evidence, draw inferences, and test coherence. Intellect provides the depth that gives reasoning meaning, drawing on prior knowledge, identifying bias and fallacy, and exercising judgment in novel situations. Intellect also creates space for creativity. It is here that the imaginative leap occurs, the moment when reasoning moves beyond mechanical procedure and becomes insight. Argumentation, in this sense, does not stand in opposition to creativity; it draws its vitality from it. Every significant advance in STEM reflects this interplay where reasoning is disciplined by intellect, and intellect animated by imagination. Without this reciprocal movement, inquiry stagnates, and innovation becomes little more than repetition.

As we move further into this chapter, the relationships among reasoning, epistemic practice, and argumentation begin to take shape as illustrated in Fig. 4.1. What initially appears as a collection of discrete skills reveals itself as an integrated

P. N. Iwuanyanwu, *Empowering STEM Thinkers Through Argumentation*, Contemporary Trends and Issues in Science Education 69,
https://doi.org/10.1007/978-3-032-27906-4_4

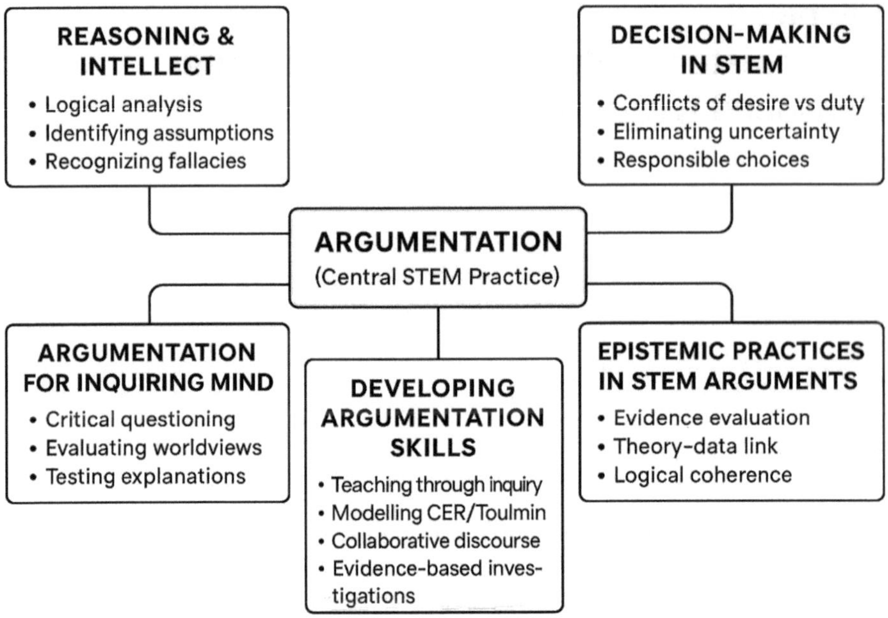

Fig. 4.1 Epistemic interdependence of reasoning, practice, and argumentation in STEM professional formation

architecture of thinking. As we move outward, we first encounter reasoning and intellect, where our inquiry begins with logical analysis, the careful surfacing of assumptions, and the disciplined recognition of fallacies. We quickly realise that these are not abstract skills reserved for philosophers; they are everyday cognitive tools we deploy when modelling a system, interpreting data, or questioning whether a result truly follows from a method. Here, we start asking ourselves: *What am I assuming? What counts as a good reason in this context?* In this way, intellect steadies our reasoning, while reasoning activates our intellect. From this grounding, our inquiry naturally extends into decision-making in STEM. We find ourselves confronting tensions between desire and duty, efficiency and responsibility, innovation and consequence. At this point, argumentation becomes more than epistemic; it becomes ethical. We begin to see that responsible STEM practice requires us to eliminate uncertainty where possible, acknowledge it where necessary, and justify our choices transparently. We are compelled to ask not only *Can this be done?* but also *Should it be done, and on what grounds?*

As we continue along this path, our attention shifts toward argumentation as a disposition of an inquiring mind. Here, inquiry is sustained through critical questioning, the evaluation of competing worldviews, and the testing of explanations against evidence and counter-evidence. We recognise that progress in STEM rarely comes from defending a single perspective uncritically (Dönmez et al., 2022). Instead, it emerges when we allow our ideas to be challenged, refined, or even abandoned. In this space, argumentation trains us to remain intellectually agile rather than ideologically rigid. Our journey then brings us into the domain of epistemic practices in STEM arguments, where evidence is no longer taken at face value but is weighed, interpreted, and linked coherently to theory. We come to appreciate that strong arguments depend on how well data are connected to explanatory frameworks and how consistently claims cohere with established knowledge. At this stage, we begin to see argumentation as a quality control mechanism for STEM knowledge itself, guarding against error, bias, and unwarranted certainty. These strands converge in the development of argumentation skills, where inquiry becomes pedagogical and collaborative. Teaching through inquiry, modelling structures such as Toulmin's scheme or Contiguity Argumentation Theory (Ogunniyi, 2007), engaging in collaborative discourse, and designing evidence-based investigations all become practical expressions of the epistemic commitments we have traced. We recognise that these practices are not only for students in classrooms but for emerging professionals who must communicate, justify, and defend decisions in complex, real-world contexts.

Seen in this way, argumentation emerges as more than a method or skill. It becomes the organising practice through which STEM inquiry remains coherent, creative, and accountable (National Research Council [NRC], 2011). What follows in this chapter builds upon this centre, tracing how epistemic practices give form to reasoning, and how reasoning, in turn, shapes responsible action and professional judgment in STEM. When we construct arguments, we move continually between deductive and inductive reasoning. Deduction carries us from general principles to specific conclusions, ensuring internal coherence. Induction moves in the opposite direction, allowing patterns to emerge from particular cases and opening new avenues of discovery. Rather than treating these forms of reasoning as opposites, we should understand them as complementary expressions of our cognitive rhythm. Deduction stabilises understanding; induction expands it. In this light, reason can be understood as movement, an active search that propels inquiry forward, while intellect represents moments of temporary rest, where understanding is achieved, however provisionally. Every act of reasoning begins and ends in understanding, much as every step in a journey begins and ends in stillness. Each movement of reason brings us closer to the intellectual horizon our curiosity seeks to reach in our study and STEM practice, even as that horizon continues to recede.

Within this framework, an argument can be understood as a line of reasoning advanced to support an intellectual stance, idea, or proposition in engagement with alternative perspectives, theories, or actions (Toulmin et al., 1979). Argumentation, then, is the structured exercise of both practical and theoretical reason (Iwuanyanwu, 2017). It responds to a fundamental human question: *What should we know, and how should we come to know it?* Through argumentation, we place reason and intellect in systematic dialogue with uncertainty. We explore perspectives, interrogate assumptions, and arrive at conclusions that are reasoned rather than absolute. In doing so, we invite critique, refine our thinking, and contribute to collective understanding. Argumentation thus becomes a vehicle not only for persuasion, but for intellectual growth. In STEM contexts, argumentation carries an additional social and ethical dimension. It is the mechanism through which misconceptions are challenged, claims are scrutinised, and evidence-based decision-making is sustained. When scientists and engineers articulate claims grounded in logic and supported by empirical evidence, they do more than strengthen their conclusions; they model an epistemic culture in which critique is welcomed and fallibility acknowledged.

Such practices are essential to maintaining the integrity of research and enabling collaboration across disciplinary boundaries. Through dialectical engagement, gaps in understanding are identified, theories are tested, and new hypotheses are proposed. Argumentation, in this sense, is not merely individual reasoning made public; it is a collective epistemic practice that advances both theory and application. As I have argued elsewhere (Iwuanyanwu, 2023), sustained engagement in argumentation cultivates epistemic virtues, rigor, openness, reflexivity, that are indispensable to both scientific inquiry and democratic life. Recent scholarship has reinforced the pedagogical importance of argumentation in STEM education. A growing body of research demonstrates that when students are engaged in structured argumentation, their conceptual understanding deepens and their reasoning becomes more sophisticated (e.g., Erduran & Park, 2023; Koichu et al., 2022; Ogunniyi & Iwuanyanwu, 2024; Tarekegn et al., 2022; Wright et al., 2025). Rather than memorising content, students learn to participate in the epistemic practices of science itself.

When students are asked to construct, justify, and critique claims using evidence, they develop the capacity to weigh alternatives, evaluate uncertainty, and make judgments that integrate cognitive and moral reasoning. These findings align with earlier work (Iwuanyanwu, 2022; Wright et al., 2025) showing that argumentation transforms STEM classrooms into spaces of inquiry, where knowledge is constructed, negotiated, and justified rather than transmitted. From this perspective, cultivating argumentation in STEM is not simply an instructional technique; it is a moral and professional commitment (Ogunniyi & Iwuanyanwu, 2024). It requires attention to how practical reasoning, concerned with action and ethics, interacts with theoretical reasoning, concerned with explanation and truth. By guiding students to reason across both dimensions, educators prepare them not only to know, but to judge; not only to design, but to discern. In a world increasingly shaped by frontier technologies, this capacity is indispensable. Argumentation functions here as both method and compass, a means of inquiry that ensures intellect remains human, reflective, and responsible in the face of accelerating automation and technological power (Osborne, 2019).

4.2 Developing Decision-Making Skills Through Argumentation

Decision-making and choice-making are not auxiliary competencies in STEM practice; they are essential cognitive and moral capacities that must be cultivated deliberately in the twenty-first century. The world in which STEM professionals operate is defined by crises that test the limits of human reason: climate change, ecological degradation, technological automation, and socio-economic instability. Many of these challenges are direct consequences of scientific, technological, and industrial activity (Aslam et al., 2023). Their effects are visible in floods, droughts, heat waves, wildfires, and the erosion of ecosystems and livelihoods, as well as in less visible but equally consequential forms such as contaminated water and soil, the spread of disease, and rising unemployment. These conditions are not isolated events. They are manifestations of what we describe as socio-STEM entanglements, complex, human-induced phenomena that demand collective, reasoned, and ethically responsible responses.

Within this context, a transformation in STEM education becomes unavoidable. When STEM learning privileges the transmission of theoretical knowledge without cultivating decision-making grounded in argumentation, we risk preparing graduates who can model systems but struggle to reason through their moral and practical consequences. The conditions confronting contemporary societies compel us to teach STEM differently. We must shift from a mechanistic emphasis on content delivery toward the development of reasoned judgment. Learners must acquire the capacity not only to know, but to decide; not only to analyse data, but to discern which actions are justifiable, sustainable, and humane. Practical reasoning plays a central role in this transformation. Unlike theoretical reasoning, which aims to establish what is true, practical reasoning guides what ought to be done. It generates imperatives rather than propositions, grounding decisions in reasons that are both epistemic and ethical. Yet in STEM education and practice, decision-making is often conflated with choice-making, as though the two were interchangeable. They are not. A choice involves perception and preference. According to Abney et al. (2014), the interplay of choice and preference includes the evaluation and selection of alternatives based on appeal, advantage, habit, or inclination. A decision, by contrast, involves judgment. It is informed by reasons, practical, theoretical, or both, and culminates in a commitment to one course of action as more rationally or ethically defensible than others.

Decision-making therefore unfolds as a movement between openness and commitment. When confronted with a decision, we often lean tentatively toward one direction while preserving the possibility of revision. We may arrive at a conclusion intellectually while postponing action until conviction, necessity, or moral resolve compels us forward. This interplay of inclination and hesitation reveals the reflective tension inherent in reasoning itself. Whether in scientific inquiry or professional practice, deciding requires an alignment of reason, value, and action. As illustrated in Fig. 4.2, decision-making is not linear. It involves recursive deliberation shaped by internal dispositions and external pressures, including personal values, social norms, disciplinary conventions, and situational constraints.

Building on this understanding, Fig. 4.2 invites us, as STEM practitioners and future professionals, to examine more closely how we move from deliberation to decisive action. In our laboratories, classrooms, design studios, clinics, and field sites, our work rarely proceeds in a simple straight line. We begin along parallel tracks of possibility, collecting data, testing assumptions, consulting standards,

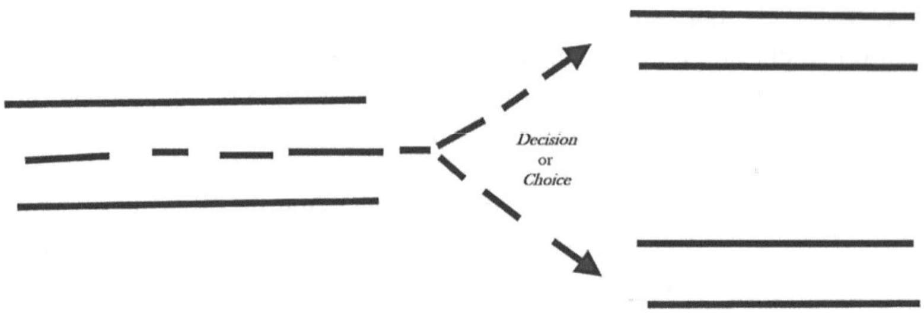

Fig. 4.2 Making a decision or choice

weighing feasibility against innovation. At this stage, we remain intentionally open. We generate models, compare alternatives, simulate outcomes. We lean toward promising directions while preserving the option to revise. As we advance, however, our inquiry converges toward a critical juncture, *decision or choice*. For the practicing engineer, this may be the moment of selecting a design under constraints of cost, safety, and sustainability. For the science educator, it may involve choosing an instructional strategy that balances curriculum demands with learners' diverse needs. For students preparing to enter STEM professions, it may be the decision to commit to a methodological approach, a research question, or an ethical stance in practice. Here, openness encounters responsibility. Evidence must align with professional judgment; innovation must remain accountable to standards, ethics, and societal impact.

From this point, our pathway diverges. One direction may represent incremental refinement, optimizing within established paradigms. Another may signal transformative innovation, venturing into emerging technologies, interdisciplinary collaboration, or novel pedagogical approaches. Each path carries consequences. Our decisions shape not only technical outcomes but also the learning cultures we cultivate, the problems we prioritize, and the communities we serve. Importantly, the movement depicted here reminds us that decision-making in STEM is recursive rather than rigid. Even after commitment, we monitor, evaluate, and iterate. Data may redirect us. User feedback may prompt redesign. Classroom evidence may require pedagogical adjustment. Thus, commitment does not silence reflection; it disciplines it.

For today's and tomorrow's STEM innovators, this movement between openness and commitment is foundational. Creativity in STEM is not unbounded improvisation; it is disciplined imagination grounded in evidence, ethical awareness, and professional standards. As practitioners, we model this process for our students. As emerging professionals, students learn that thoughtful hesitation is not weakness but intellectual rigor, and that decisive action, when informed by reason and value, becomes the engine of advancement. In tracing this movement together, we are reminded that every significant breakthrough, whether in scientific discovery, technological design, or STEM education reform, emerges from this disciplined interplay. The future of STEM learning and practice depends on our capacity to deliberate deeply, decide responsibly, and innovate courageously.

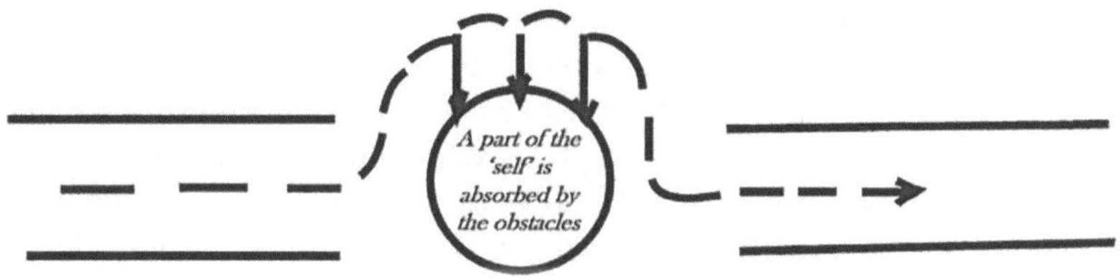

Fig. 4.3 Relative obstacles in decision-making

Better still, sound decision-making demands attentiveness to both theoretical and practical reasons. Before acting, we must analyse potential outcomes, weigh benefits against risks, and assess the worthiness of the proposed course. Such deliberation enables alignment between cognitive coherence and moral integrity. Whether selecting a research design, developing a technological system, or shaping environmental policy, decision-making reflects our intellectual posture toward uncertainty. To decide well is to act with clarity about what we value and why. In this sense, decision-making functions as an epistemic practice, one that refines judgment and extends understanding.

At the same time, decision-making is rarely transparent or effortless. It is frequently obscured by inertia, fear, complacency, or the subtle comforts of conformity. A persistent challenge lies in recognising when apparent deliberation is, in fact, avoidance. As suggested in Fig. 4.3, the gap between decision channels represents the cognitive and emotional space in which uncertainty and hesitation dwell. This space should not be dismissed as resistance alone. It is a site of ethical struggle, where intellect contends with will. To decide is to bridge this gap, to move from contemplation to commitment. External factors further complicate this movement. Obstacles, whether informational, structural, or psychological, can distort reasoning and obscure judgment. These barriers transform decision-making into a challenge requiring both intellectual clarity and moral ingenuity. It is important, however, to distinguish decision-making from problem-solving. While problem-solving aims to resolve a difficulty, decision-making concerns commitment to a course of action even when problems remain complex, uncertain, or partially unresolved. In practice, each decision reshapes the problem space itself, altering the conditions under which further inquiry proceeds.

For this reason, developing decision-making through argumentation is indispensable in STEM education. Argumentation compels us to articulate reasons, confront counterarguments, and deliberate among competing claims. It transforms decision-making from an impulsive or habitual act into a reasoned, dialogic process that integrates epistemic rigor with ethical responsibility. Through argumentation, learners cultivate essential epistemic virtues. These include intellectual humility, responsiveness to evidence, and the capacity to act wisely under conditions of uncertainty. The aims of STEM education therefore extend beyond producing efficient problem-solvers or innovative designers (Mahmood et al., 2025). We must prepare learners to reason well, to decide with foresight, and to act with conscience. Argumentation functions as the pedagogical medium through which knowledge becomes judgment and intellect matures into wisdom. Figure 4.3 compels us, as STEM practitioners and future professionals, to confront a reality we know well, namely, that even the most carefully reasoned decisions encounter resistance. We have affirmed that argumentation disciplines our thinking, requiring us to justify claims, interrogate assumptions, and deliberate responsibly. Yet when we move from seminar rooms and structured design reviews into laboratories, classrooms, industries, and communities, our reasoning meets relative obstacles.

We may begin with methodological clarity, technical competence, and ethical intention. Our progress appears steady, guided by data, standards, and collaborative dialogue. But in practice, the path bends. Institutional policies constrain innovation. Funding limitations shape research trajectories. Safety regulations restrict design options. Cultural expectations influence classroom implementation. Emerging professionals may encounter imposter feelings, time pressures, or conflicting supervisory advice. In such moments, a portion of the "self" is absorbed. What is absorbed? Often it is our initial certainty. Sometimes it is our preferred solution. At times, it is the comfort of operating within familiar paradigms. Obstacles demand cognitive, emotional, and ethical expenditure. They require us to relinquish simplifications and confront complexity. For the engineer, this may mean redesigning under stricter sustainability requirements. For the educator, it may involve adapting instruction to diverse learners while meeting accountability standards. For the researcher, it may require revising hypotheses when evidence resists expectation. Yet the movement does not terminate at obstruction. It re-emerges, refined. This is where argumentation proves indispensable. Because we have cultivated intellectual humility, responsiveness to evidence, and disciplined deliberation, we are not immobilized by resistance. Instead, we recalibrate. We iterate. We transform constraint into catalyst.

For today's and tomorrow's STEM innovators, obstacles are not anomalies; they are generative conditions. Many of the most significant advancements, green technologies, inclusive pedagogies, interdisciplinary breakthroughs, have emerged precisely because practitioners refused to retreat when confronted with constraint. Rather than allowing obstacles to erode professional identity, they allowed them to sharpen judgment and deepen responsibility. As a scholarly community committed to advancing STEM learning and practice, we must therefore prepare students not merely to solve idealized problems but to navigate friction intelligently. We must model how to reason under pressure, how to defend decisions with evidence and ethical clarity, and how to adapt without compromising integrity. In doing so, we recognize that obstacles do not simply test our expertise; they refine it. They absorb what is unexamined and strengthen what is principled. The enduring question before us, whether as experienced practitioners or emerging professionals, is this: *When our reasoning encounters resistance, will we experience diminishment, or will we allow that encounter to mature our innovation, our judgment, and our contribution to the evolving future of STEM?*

As can be seen in Fig. 4.3, obstacles are a defining feature of decision-making. These are not merely external impediments but epistemic and existential challenges that test reasoning, patience, and moral resolve. Some obstacles are visible and measurable, constraints of time, resources, or technical feasibility. Others are subtle

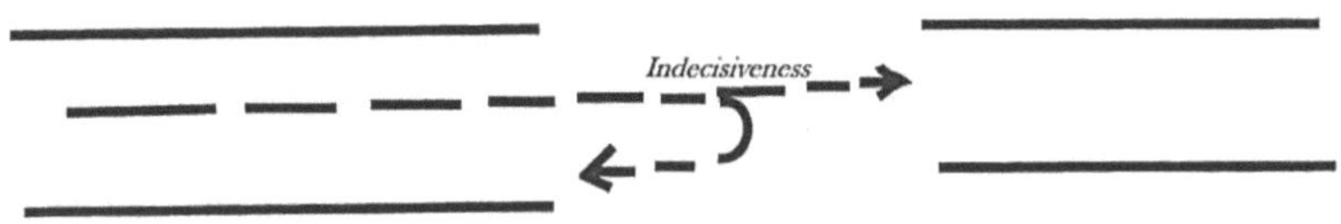

Fig. 4.4 Inability to make a decision or choice

and elusive, arising from uncertainty, fear, conflicting values, or incomplete evidence. Often, what appears most formidable is not the obstacle itself but the magnitude we assign to it in anticipation. Regardless of form, obstacles draw upon our cognitive and emotional resources, demanding clarity about what truly matters in the decision at hand. Responsible decision-making requires careful examination of the conditions that make judgment possible. This include the quality of evidence, the coherence of reasoning, and the soundness of justification. When obstacles prove insurmountable, intellectual honesty demands reassessment rather than persistence for its own sake. Decisions that cannot withstand scrutiny grounded in evidence, reason, and explanation lack the stability required for real-world complexity.

To reason well, therefore, is to decide well. Every decision generates consequences that ripple across personal, social, and ecological systems. In STEM contexts, this moral dimension cannot be treated as secondary. Decisions about energy systems, technological design, or environmental intervention shape the conditions of human and planetary well-being. Avoiding a decision does not eliminate responsibility. Inaction is itself a form of choice, carrying consequences no less significant than deliberate action (see Fig. 4.4). When learners and practitioners understand decision-making as both an epistemic and ethical act, they begin to connect cognitive skill with human consequence. They become better equipped to engage complex issues, climate change, sustainability, technological risk, and socio-technical inequity, through a lens that balances evidence with moral discernment.

By foregrounding the interplay among evidence, reasoning, and explanation, STEM education cultivates not only technical competence but intellectual responsibility. Educators and researchers bear a particular responsibility to model these epistemic virtues. Pedagogical practices must consistently place valid evidence, sound reasoning, and plausible explanation at the centre of STEM inquiry. Only then can we prepare learners not merely to solve problems, but to decide wisely, to act as informed agents committed to sustainable and humane futures. Argumentation, as this chapter demonstrates, is not simply a method of persuasion. It is the practice through which decision-making becomes an expression of intellect, responsibility, and hope.

If Fig. 4.3 depicts resistance encountered, Fig. 4.4 confronts us, as STEM practitioners and future professionals, with a condition that can quietly undermine innovation, the inability to decide. We may begin with data collected, models constructed, alternatives compared. Our analysis appears rigorous, our reasoning disciplined. Yet at a critical moment, forward motion falters. Instead of converging toward commitment, our thinking loops. We revisit simulations, request additional data, delay implementation, or defer responsibility. In research settings, this may appear as perpetual refinement of methodology without publication. In engineering practice, it may manifest as extended feasibility studies without deployment. In STEM classrooms, it may surface as continual curriculum redesign without instructional execution. We remain analytically active, yet strategically stalled. Indecisiveness in STEM is rarely born of laziness; it often arises from high standards, risk awareness, or fear of unintended consequences. We seek optimal solutions in environments that only permit timely ones. We desire certainty in systems defined by complexity. However, innovation demands more than analysis, it requires courageous, evidence-informed commitment.

For emerging professionals, this moment is formative. The transition from student to practitioner involves learning when evidence is sufficient, when iteration must occur in practice rather than theory, and when responsibility requires action despite residual uncertainty. Intellectual humility must be balanced with professional resolve. The looping movement before us therefore serves as caution and challenge. If we remain indefinitely within recursive deliberation, opportunities pass, technologies stagnate, and educational reforms remain conceptual. Yet if we cultivate disciplined judgment, grounded in argumentation, ethical awareness, and responsiveness to evidence, we learn to convert analysis into advancement.

In advancing STEM learning and practice, we must model for our students that decisive action is not recklessness but responsibility exercised under uncertainty. The future of sustainable technologies, equitable STEM education, and transformative discovery depends not only on how well we analyze problems, but on whether we are prepared to choose, and to act, when the moment demands it. The enduring question for us, then, is this: *Are we refining our thinking toward commitment, or are we circulating indefinitely within it?*

Conversely, indecisiveness is often misunderstood as weakness, yet it frequently arises from the presence of multiple plausible alternatives rather than from ignorance. When each option contains both promise and limitation, hesitation reflects a desire for certainty in conditions where certainty is unavailable. Kant's account of practical reasoning helps clarify this tension, distinguishing conflicts among inclinations from deeper conflicts between inclination and duty. In STEM practice, such conflicts are unavoidable. Personal ambition, institutional pressure, and societal demand often collide with ethical responsibility. The capacity to navigate this moral–epistemic tension is among the highest expressions of professional judgment. Here, argumentation becomes indispensable. By articulating reasons, testing claims, and negotiating conflicts within a rational framework, indecision is transformed into deliberation. Uncertainty becomes not an obstacle but a condition for intellectual growth. Through argumentation, decision-making becomes disciplined inquiry, a fusion of evidence, reasoning, and ethical imagination. In this way, indecision ceases to paralyse and instead becomes possibility, the space in which judgment is formed and wisdom begins to emerge.

4.3 Epistemic Practices in the Use of Arguments in STEM

When we reflect on the nature of epistemic practice in STEM, we begin to see more clearly that argumentation is not an accessory to technical work. It is the disciplined activity through which our knowledge claims acquire legitimacy, coherence, and direction. As practitioners and future professionals, we do not argue merely to defend positions; we argue to test, refine, and extend understanding. Argumentation is therefore not persuasion in the ordinary sense. It is participation in a community committed to rational accountability. In our laboratories, design studios, classrooms, and research environments, arguments emerge wherever claims are made, models are proposed, data are interpreted, or systems are evaluated. Each claim invites a question: *On what grounds?* Each explanation requires linkage between evidence and theory. Each innovation demands justification beyond enthusiasm. Here, argumentation becomes the living structure of epistemic integrity. It requires that we move beyond intuition and preference toward transparent reasoning, systematic evidence use, and logical coherence.

When we engage in argumentation responsibly, we make our thinking visible. We expose assumptions, articulate warrants, acknowledge limitations, and anticipate counterarguments. In doing so, we transform private cognition into public reasoning. The strength of our conclusions does not rest on conviction or authority, but on the quality of the inferential bridge between data and claim. This is the discipline that protects STEM from dogmatism and shields innovation from becoming speculation. Within the broader STEM community, our arguments are never solitary possessions. They enter what we might call a republic of reason, a space where critique is not hostility but contribution. Peer review, replication studies, design validation processes, and interdisciplinary panels are institutional expressions of this epistemic culture. When our arguments withstand scrutiny, they gain robustness. When they fail, we do not experience defeat but refinement. In this way, argumentation becomes the quality-control mechanism of STEM knowledge itself.

Importantly, the epistemic practice of argumentation extends beyond disciplinary silos. The defining challenges of our era, climate instability, renewable energy transitions, bioengineering ethics, artificial intelligence governance, socio-technical inequities, require integrated reasoning. No single discipline can exhaustively account for their complexity. As practitioners, we must therefore learn to construct arguments that travel across conceptual borders. We must negotiate differences in methodology, terminology, and standards of evidence. In such spaces, argumentation becomes translational. It builds conceptual bridges where fragmentation would otherwise prevail. For STEM educators and those preparing to enter STEM professions, cultivating these epistemic practices is foundational. When we guide students to construct and critique arguments, we are not merely strengthening communication skills; we are inducting them into the epistemic culture of STEM. They begin to ask:

i) What counts as credible evidence in this context?
ii) How does this model relate to established theory?

iii) Where might bias or hidden assumptions distort interpretation?
iv) How robust is this inference under alternative conditions?

Through such questioning, learners move from procedural compliance to epistemic agency. They begin to understand that knowledge is not delivered intact but constructed through disciplined reasoning. Argumentation becomes a mirror of cognition, revealing not only what they know, but how they justify knowing. Moreover, argumentation nurtures ethical awareness. When we justify a technological design, propose an intervention, or interpret environmental data, our reasoning carries consequences. Argumentation obliges us to consider who benefits, who bears risk, and how uncertainty is acknowledged. It demands intellectual humility and responsiveness to evidence, particularly when that evidence unsettles our preferred outcomes. In this sense, epistemic rigor and ethical responsibility converge.

As we look toward the future of STEM innovation, the importance of these practices intensifies. Emerging technologies increasingly operate at scales and speeds that outpace intuitive judgment. Automation, algorithmic decision systems, and advanced modeling tools can generate outputs with remarkable efficiency. Yet without disciplined argumentation, such outputs risk becoming opaque or uncritically accepted. We must therefore ensure that human intellect remains actively engaged, interrogating assumptions, validating inferences, and preserving accountability. Ultimately, epistemic practices in the use of arguments embody the very spirit of STEM inquiry. They ensure that creativity is disciplined, that innovation is justified, and that knowledge remains open to improvement. For us as practitioners and emerging professionals, argumentation is not a peripheral academic exercise. It is the ongoing practice through which we learn to think clearly, reason responsibly, collaborate constructively, and act wisely in a world shaped profoundly by our scientific and technological decisions.

4.4 Argumentation for the Inquiring Mind

If we trace the historical arc of intellectual progress, we find that argumentation has always been the companion of inquiry. Civilizations have advanced not merely through accumulation of information, but through disciplined questioning and reasoned dispute. Across eras, thinkers have recognized that understanding deepens when ideas are examined, challenged, and justified. Yet for us in STEM, the relevance of this historical inheritance is not antiquarian; it is formative. The inquiring mind does not arise spontaneously from exposure to content. It is cultivated through engagement in structured reasoning. When we invite students, and ourselves, into argumentative discourse, we nurture a disposition that is curious yet disciplined, critical yet constructive.

An inquiring mind begins with questions that disturb complacency. Why does this pattern appear in the data? What alternative explanation might account for this anomaly? What assumptions underlie this model? Questions function as intellectual catalysts. They interrupt routine cognition and compel deeper analysis. In our classrooms and professional settings, the quality of our inquiry often depends on the quality of our questions. Argumentation sharpens the intellectual instruments with which we pursue these questions. It compels us to distinguish between inductive generalization and deductive certainty, between correlation and causation, between plausible conjecture and empirically grounded explanation. We learn to construct claims supported by evidence and governed by warrants that make explicit why the evidence is relevant. In doing so, we discipline imagination without extinguishing it.

For students preparing to enter STEM professions, this training is transformative. When students are positioned as arguers rather than passive recipients, they become critical friends of science and technology (Iwuanyanwu, 2023). They learn to test explanations rather than memorize them. They evaluate sources rather than assume credibility. They experience the intellectual satisfaction that arises when insight is earned through reasoning rather than delivered through authority. Argumentation also requires reflexivity. We must learn to recognize our own presuppositions, disciplinary biases, and value commitments. In interdisciplinary STEM contexts especially, what appears self-evident within one field may be contestable in another (Wright et al., 2025). The inquiring mind therefore cultivates epistemic humility. It acknowledges that knowledge is provisional, that conclusions remain open to revision, and that alternative perspectives may illuminate blind spots in our reasoning.

At the same time, argumentation strengthens intellectual courage. In research, innovation, and education reform, we often confront dominant paradigms. To question established frameworks responsibly requires evidence, coherence, and moral conviction. The inquiring mind is not rebellious for its own sake; it is committed to improvement. Through disciplined argumentation, critique becomes a pathway to advancement rather than disruption. For today's and future STEM innovators, the stakes are high. We operate within socio-technical systems that shape energy infrastructures, health outcomes, digital communication, and ecological stability. The complexity of these systems demands integrative reasoning. Argumentation provides the architecture through which we synthesize empirical observation, conceptual reasoning, and ethical reflection. It allows us to test not only whether something works, but whether it ought to be pursued. In educational settings, fostering argumentation means designing learning environments where dialogue is structured, evidence is central, and disagreement is treated as generative. When we model respectful critique and evidence-based defense of ideas, we cultivate classrooms that mirror authentic STEM practice. Students come to see that dissent is not disorder; it is the engine of refinement.

Ultimately, to cultivate the inquiring mind through argumentation is to reaffirm the essence of education itself. Education in STEM is not merely the transmission of formulas, procedures, or technical skills. It is the formation of individuals capable of disciplined curiosity, reasoned judgment, and ethical innovation. As practitioners and emerging professionals, we must therefore ask ourselves: Are we merely solving problems within established frames, or are we equipping ourselves and our students to question the frames themselves? When argumentation becomes habitual, inquiry becomes sustainable. Our thinking grows more agile. Our decisions become more responsible. Our innovations gain depth. In this way, argumentation does not only serve STEM; it animates it. It ensures that our pursuit of advancement remains anchored in reason, open to critique, and oriented toward the flourishing of both human and ecological communities.

4.5 The Interplay Between Valid Arguments and the Logic of Scientific Reasoning

In Sect. 4.3, we established argumentation as an epistemic practice, the disciplined activity through which knowledge claims are rendered visible, contestable, and improvable. In Sect. 4.4, we saw how the inquiring mind is formed through structured questioning and reflexive reasoning. We now move to a deeper layer, that is, the logical architecture that makes valid scientific argument possible. The power of argumentation in STEM lies not merely in its communicative function but in its epistemological necessity. To begin with, if epistemology asks, *What may we claim to know?*, and logic asks, *How must we reason for that claim to follow?*, scientific inquiry stands precisely at their intersection. In other words, when we argue scientifically, we construct inferential bridges between observation and conclusion. However, these bridges must satisfy two conditions simultaneously: (1) logical validity—the conclusion follows coherently from the premises; and (2) epistemic justification—the premises themselves are credible within a disciplined methodological framework. Without the former, reasoning becomes structurally unsound; without the latter, it lacks evidential grounding. Consequently, scientific reasoning collapses into either empty formalism or unstructured empiricism.

In view of this, argumentation provides the logical architecture through which knowledge claims acquire justification. Without it, statements remain mere assertions; with it, they become candidates for warranted belief. Accordingly, argumentation operates at the intersection of epistemology and logic. It concerns both what we may claim to know and how such claims are structurally supported. To argue scientifically, therefore, one must construct inferential pathways that connect premises to conclusions under conditions of evidential constraint. This is because the strength of a scientific claim depends not on rhetorical force but on the coherence of its logical form and the credibility of its empirical grounding. Thus, argumentation disciplines both our reasoning and our knowing.

To illuminate the interplay between logical structure and empirical validation, we turn to a classical problem that unsettles intuition and compels disciplined reasoning, the question of simultaneity in Relativity: The Special and General Theory by Albert Einstein (1920, pp. 25–27). In Fig. 4.5, we encounter a seemingly simple scene in which lightning strikes two points, P and Q, on a railway embankment, and the flashes are said to occur simultaneously. At first glance, the claim appears self-evident. If both flashes occur "at the same time," what further reasoning could be

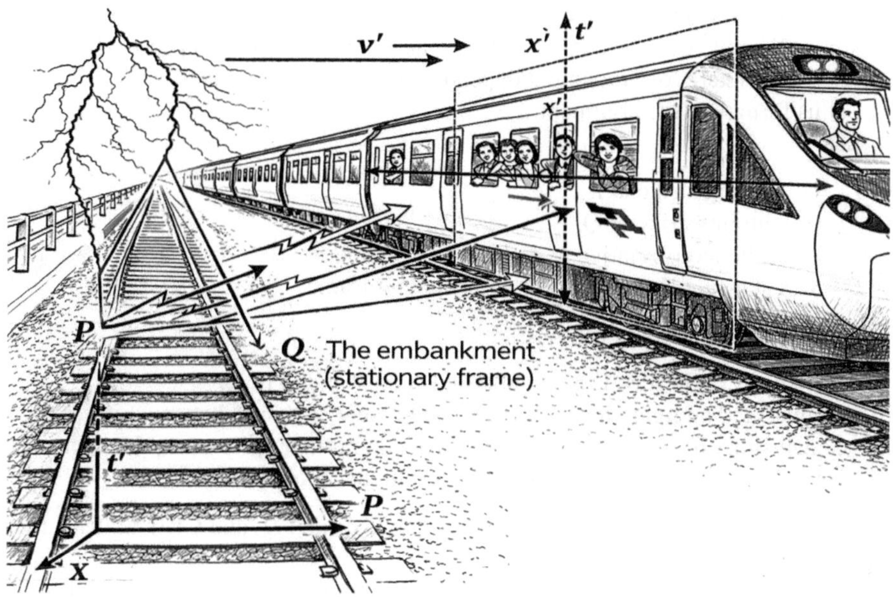

Fig. 4.5 Simultaneity across spatially distinct points

required? However, once we pause to reflect, the apparent simplicity begins to dissolve. The assertion of simultaneity is not merely descriptive; rather, it presupposes a framework of observation, measurement, and interpretation. What initially appears obvious therefore demands justification. Consequently, our inquiry intensifies as we move from intuition to analysis. Here our inquiry intensifies. We ask:

1. What does "simultaneous" mean when events occur at spatially distinct locations?
2. How do we operationalize simultaneity in measurable terms?
3. What epistemic conditions must be satisfied before we may affirm such a claim?

We begin by imagining light rays emitted from P and Q traveling toward a midpoint X on the embankment. If those rays converge at X at the same moment, observers stationed there judge the strikes simultaneous relative to the embankment. This spatial configuration is represented in Fig. 4.5.

Here we see more than geometry; we see an epistemic arrangement. The midpoint X becomes more than a location; it functions as an epistemic checkpoint. It reminds us that perception alone does not settle questions of temporal order. Evidence must be interpreted within a defined system of coordinates. As practitioners and future professionals, we learn here that even the most straightforward claims conceal conceptual commitments. Argumentation begins precisely where intuition feels sufficient. We begin with an observer positioned at midpoint X between P and Q along the embankment. The midpoint X functions as a site of observational symmetry. It embodies an implicit premise, that the speed of light is constant and isotropic relative to the embankment frame. Stated otherwise, light rays from P and Q converge at X. At first glance, our reasoning appears straightforward. Yet even here, our judgment depends upon implicit premises.

Let us formalize the argument:

Let:

- L_P = Lightning strike at P
- L_Q = Lightning strike at Q
- S_E = Simultaneous relative to embankment frame
- C = Constancy of light speed in embankment frame

The inference proceeds as follows:

i) Premise 1: Light travels at equal speed from P and Q to X.
ii) Premise 2: The light signals from P and Q meet at X simultaneously.
iii) Conclusion: If equal-speed signals arrive simultaneously at equidistant midpoint, then the events were simultaneous in that frame

Therefore: $(L_P \wedge L_Q \wedge C) \rightarrow S_E$ (we reason that if L_P and L_Q and C are all true, then S_E follows).

This is not mere description; it is a deductive–empirical synthesis. Logical structure (premises and conclusion) is inseparable from empirical assumption (light-speed constancy). The epistemic legitimacy of the conclusion depends on the validity of both. In other words, when all three premises (L_P, L_Q, and C) are satisfied simultaneously, then the conclusion S_E must follow. Importantly, this does not say that S_E guarantees the premises. It only states that the premises are sufficient for the conclusion. Here we already see epistemological layering. Premise (1) is not empirically self-evident; it rests upon a theoretical postulate (in Einstein's framework, invariance of light speed). The conclusion follows deductively, but only if the premises are defensible. Thus, logical validity depends on epistemic credibility. Structurally, this expression $(L_P \wedge L_Q \wedge C) \rightarrow S_E$ expresses a conditional scientific claim in three stages: (i) A conclusion (S_E) depends on a conjunction of conditions, (ii) If any one of the premises is false, the implication does not hold, and (iii) The strength of the conclusion depends on the truth of all contributing premises. In philosophical terms, it shows how a scientific claim is not asserted independently but is logically derived from specified evidential conditions. Already we see how argumentation transforms intuition into disciplined reasoning. The simplicity of the claim conceals theoretical commitments regarding space, time, and measurement. Without articulating these commitments, the claim remains epistemically opaque.

Our inquiry deepens when motion enters the scenario. Suppose a train moves with constant velocity (v) relative to the embankment. Observers inside the train form their own inertial reference frame. Within this moving system, spatial and temporal judgments are coordinated differently. Now we must ask: if light from P and Q travels toward an observer seated at the train's midpoint, will those signals arrive simultaneously? The answer is no longer obvious. The simultaneity judgment is frame-indexed. The same physical events may yield different temporal judgments depending on the reference body within which they are evaluated. This moment is

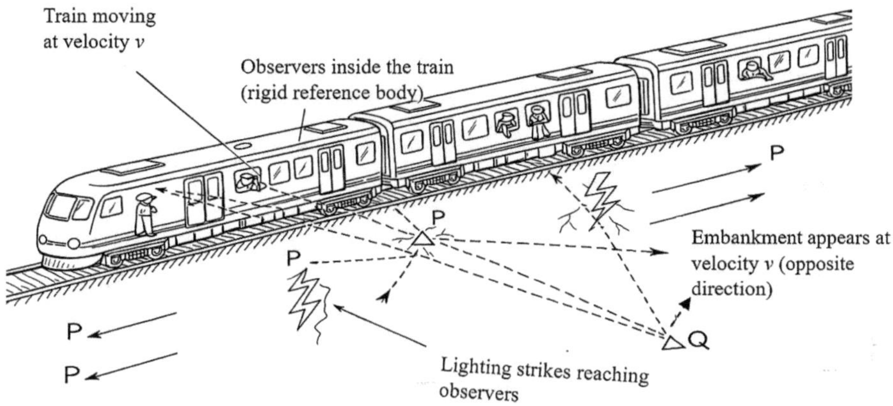

Fig. 4.6 Different observations of lightning strikes at P and Q from two inertial reference frames

philosophically decisive. We encounter what Kant would describe as the conditions of possible experience. Temporal ordering is not given "in itself" but structured through the forms under which observers synthesize experience.

As we travel mentally with these observers, we experience a shift in epistemic orientation. What was simultaneous in one frame may not be simultaneous in another. The divergence is not a contradiction of reality but a reminder that claims about time and space are meaningful only within a defined conceptual and empirical structure. Here, argumentation becomes indispensable. We cannot rely on intuition; we must articulate premises, clarify assumptions, and examine coherence. We are reminded that observation is never raw. It is always mediated through conceptual and spatiotemporal frameworks. Argumentation now becomes necessary to adjudicate between frame-dependent claims, not by eliminating one, but by clarifying the conditions under which each holds.

To make progress, we construct a virtual connection, symbolized as X, that links P and Q within a unified explanatory framework. X is not a physical wire joining the two lightning strikes. It is the mediating principle that renders them intelligible together, that is, a shared coordinate system, a law of physics, a synchronization convention. Without such a connection, P and Q remain isolated events. With it, they become relationally interpretable. This epistemic divergence is visualized in Fig. 4.6.

Now we encounter a philosophical turning point. Observers on the embankment and observers on the train do not merely disagree psychologically; they operate within distinct coordinate systems that structure measurement itself. This does not imply relativism in the sense of arbitrariness. Rather, it demonstrates that scientific claims are also frame-indexed. The epistemic lesson is profound, that is, observation is theory-laden. Judgments of simultaneity depend on synchronization conventions, spatial metrics, and assumptions about light propagation. Identical physical events can yield divergent temporal orderings when interpreted within distinct but internally coherent systems. Here, logic alone cannot resolve the question. We must

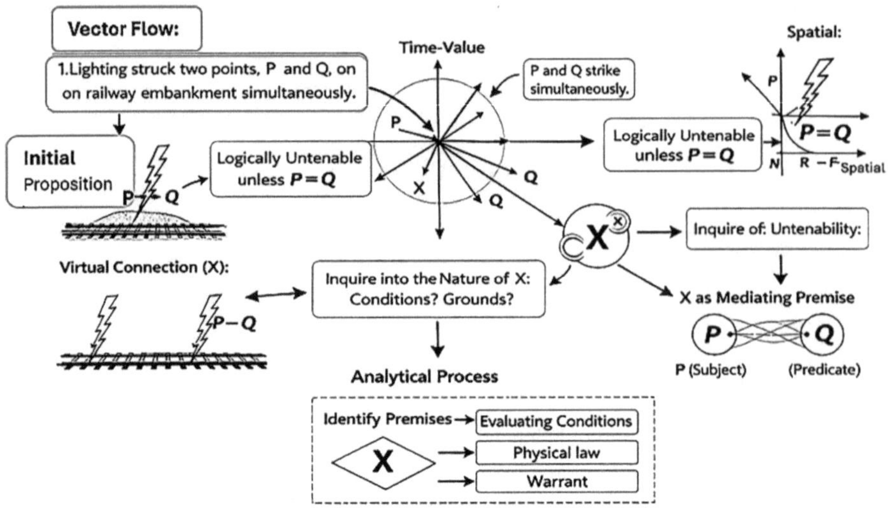

Fig. 4.7 Analytical framework for evaluating simultaneity via virtual connection X

examine the conditions under which each frame constructs temporal relations. Argumentation becomes the medium through which these conditions are articulated and compared.

To proceed analytically, we introduce a virtual connection, symbolized as X, that unites P and Q within a shared explanatory structure. This analytical move is formalized in Fig. 4.7.

X does not represent a physical object; it represents the mediating premise that allows P and Q to be conceptually linked. In logical terms, X functions as the warrant connecting subject and predicate. Without such a mediating principle, P and Q remain epistemically isolated. Formally, let:

X = Synchronization principle within defined frame
P = Event at P
Q = Event at Q

Then, the reasoning structure now becomes explicit:

- Major Premise: Events connected through synchronization principle X share temporal equivalence.
- Minor Premise: Lightning strikes at P and Q are connected through X.
- Conclusion: Therefore, P and Q are simultaneous (within that frame).

Here we witness the fusion of syllogistic logic with empirical method inferring that all events linked by X are temporally equivalent. P and Q are linked by X. Therefore, P and Q are temporally equivalent. But X is not arbitrary. It represents a synthetic principle that unites spatially distinct phenomena under a rule. In Kantian terms, it functions as a regulative structure allowing coherent experience. X may represent a physical law (e.g., invariance of light speed), a coordinate convention, or a causal

Table 4.1 Linking data, warrant, and claim for Einstein's lightning thought experiment

Toulmin element	Content
Claim	Lightning strikes at P and Q were simultaneous.
Data	Light signals from P and Q reached midpoint X simultaneously.
Warrant (X)	Equal-speed light arriving simultaneously at an equidistant midpoint indicates simultaneity.
Backing	Empirical invariance of light speed (validated experimentally).
Qualifier	Simultaneous within the embankment frame.
Rebuttal	Not simultaneous relative to moving train frame.

mechanism. The validity of the conclusion depends on whether X is defensible both logically and empirically. This illustrates a foundational epistemological insight that scientific explanation requires mediation. Phenomena do not justify themselves; they are rendered intelligible through structured inference.

Next, we turn to the internal coherence of the argument itself, applying Toulmin's framework through the integration of Warrant, Backing, and Rebuttal. First, premises must support conclusions without contradiction, ensuring that the reasoning is internally consistent. At the same time, these premises must correspond to measurable reality, grounding the argument in empirical evidence. Thus, the strength of an argument depends on this dual evaluation, both logical consistency and empirical adequacy. Here we examine:

i) Are the premises logically ordered?
ii) Is the mediating principle X empirically justified?
iii) Does the conclusion follow necessarily or probabilistically?

Let us now map the reasoning explicitly through Toulmin's model. In Table 4.1, we trace how our observations of light signals (Data) lead us to the claim that the lightning strikes at P and Q were simultaneous. We see how the Warrant links these observations to the claim, supported by the empirical Backing of light's invariant speed. At the same time, we remain attentive to the frame of reference, captured in the Qualifier, and recognize the limits of our conclusion when viewed from a moving train, as noted in the Rebuttal. In following this structure, we are guided step by step from observation to reasoned judgment, experiencing the disciplined interplay of logic and empirical evidence that underpins scientific argumentation.

Now, we see argumentation as a richly layered process. Argumentation at this level becomes both analytic and synthetic. Logic secures structural validity; empirical investigation secures material truth. Scientific reasoning fails when either dimension collapses, when coherence lacks evidence, or evidence lacks coherence. In Table 4.1, we are not simply stating a claim; instead, we provide the inferential linkage that connects our evidence to the conclusion (Warrant), we justify that linkage through supporting principles or empirical evidence (Backing), we clarify the limits of our claim (Qualifier), and we anticipate possible counter-positions (Rebuttal). In following this structure together, we transform a simple assertion into a disciplined epistemic practice, experiencing how reasoning and evidence work hand in hand to guide our understanding. Importantly, validity extends beyond correspondence with data. It concerns credibility in our inquiry, where we attend to the methodological integrity of our measurements, the reliability of the instruments we use, and our openness to falsification. In this way, we see that establishing knowledge is not just about collecting data, but about ensuring that every step of our process can be trusted and critically examined. Argumentation disciplines us to test not only conclusions but also the conditions under which they are formed.

On this basis, we proceed through a complete cycle of reasoning. We inquire into the nature of a phenomenon, analyze its premises, conditions, and relationships, and construct an argument that links evidence to conclusion. Along the way, we attend to the epistemic validity of the logical structure, verify empirical adequacy through scientific methods, and integrate these layers to ensure methodological rigor. In doing so, we recognize that validity is multi-layered. Formal validity ensures that conclusions follow from premises. Empirical adequacy requires correspondence with measurable reality. Methodological integrity guarantees replicable procedures, while conceptual coherence demands precise definitions. This reflexivity is not a flaw but an unavoidable feature of rational inquiry. Logic functions both as foundation and instrument. Recognizing this dual role protects us from absolutism and cultivates intellectual humility.

Next, we turn to epistemic provisionality and the ethics of justification. The simultaneity inquiry reveals something even more profound. Time and space are not merely empirical findings; they are conditions structuring experience. Kant argued that the mind synthesizes sensory input through forms of intuition and categories of understanding. Scientific reasoning operates within such structures. When we introduce X, we are not discovering a metaphysical wire between P and Q. We are instituting a rule under which phenomena are rendered intelligible. Scientific laws function similarly. They are not visible objects but principles organizing appearances. Argumentation becomes the disciplined act through which we justify these organizing principles. Moreover, practical reason enters when we apply such knowledge technologically. Decisions based on relativistic synchronization affect engineering systems (e.g., GPS satellite calibration). Here epistemology meets responsibility. We must also acknowledge that logical principles themselves are justified through reason. This introduces a reflective circularity:

i) We use logic to justify scientific reasoning.
ii) We use scientific reasoning to validate empirical claims.
iii) We use empirical success to reinforce confidence in logic.

This circularity is not vicious; it is constitutive of rational inquiry. It reflects what Kant called the self-legislation of reason. Logic is both tool and ground. Recognizing this prevents dogmatism. It reminds us that even our most trusted structures require ongoing justification. That is to say, no scientific claim is immune to revision. Every conclusion remains provisional, open to refinement through new evidence or more comprehensive reasoning. Argumentation ensures that knowledge remains self-correcting rather than dogmatic. For STEM practitioners and emerging professionals, the lessons are clear. No claim is self-justifying; every conclusion depends on articulated premises, every warrant requires careful backing, every framework demands qualification, and every decision carries both epistemic and ethical weight.

In an era shaped by AI systems, climate modeling, bioengineering, and algorithmic governance, the capacity to evaluate mediating principles has become indispensable. Indeed, every model carries implicit warrants, every dataset presupposes conventions of synchronization, and every technological deployment relies on underlying inferential structures. For emerging professionals, this means that the question is no longer simply Does it work? Instead, we must continually ask: What premises underlie this claim? What warrants connect evidence to conclusion? Under what frame does validity hold? Where might revision be required? Moreover, the relevance of disciplined argumentation extends far beyond acquiring technical knowledge. Whether we are analyzing AI outputs, evaluating climate projections, assessing biomedical trials, or designing engineering simulations, we must ask: What is the mediating principle? What are the unstated assumptions? What frame structures this observation? Under what conditions does the claim hold? By confronting these questions, we recognize that argumentation is not merely an accessory skill; it is an essential practice of epistemic stewardship, guiding both our reasoning and our responsibility as knowledge-makers. In a world shaped by rapidly evolving technologies, this capacity is indispensable. The ability to construct and evaluate valid arguments is not peripheral to STEM literacy; it lies at its very core.

This perspective prepares us to see, as the lightning example illustrates, that scientific reasoning is neither purely empirical nor purely logical. Rather, it is synthetic, uniting data and structure through disciplined argumentation. Logic provides form, method provides grounding, and epistemology provides justification, but it is argumentation that integrates them into a coherent whole. When we cultivate this interplay in STEM education, we do more than train competent technicians. We nurture reflective reasoners, professionals capable of bridging observation and conceptual clarity, balancing innovation with responsibility, and navigating certainty with humility. In this disciplined integration, scientific reasoning becomes not only powerful, but also wise.

At a deeper level, the simultaneity problem reveals a convergence of philosophical traditions. First, Kant reminds us that knowledge depends on conditions of possible experience, showing that temporal and spatial judgments are structured through forms of intuition and conceptual synthesis. Next, Einstein demonstrates that these structures are not naïvely absolute; rather, they are frame-dependent and must be rigorously redefined under empirical constraint. Finally, Toulmin provides the practical architecture of scientific justification, claim, data, warrant, backing, qualifier, and rebuttal, ensuring that our reasoning remains explicit, accountable, and traceable. Taken together, these perspectives convey a single insight, that scientific knowledge is not merely discovered; it is actively constructed through disciplined mediation between empirical observation and conceptual structure. In this light, argumentation emerges as the practice that performs this mediation, guiding us step by step from evidence to reasoned, justified understanding, and grounding the epistemic responsibilities of STEM practice in both logic and observation.

4.6 Setting the Stage for Developing Argumentation Skills

If, as we established in Sects. 4.3 and 4.4, argumentation functions as the epistemic lifeblood of STEM inquiry, then its cultivation cannot be incidental. It must be intentional, structured, and sustained. The movement traced earlier, from reasoning and intellect (Fig. 4.1) to epistemic practice, now becomes pedagogical responsibility. The persistence of the transmission model in STEM education reflects an understandable concern for coverage and efficiency (Tytler, 2020). Yet coverage without cognitive engagement produces fragile understanding. Information delivered intact rarely becomes durable insight. If knowledge is to become what Bruner metaphorically described as intellectual "gold," it must be mined through inquiry, not deposited through instruction. We therefore confront pedagogical tensions: How do we deliberately cultivate argumentation as the epistemic lifeblood of STEM inquiry? How do we move from acknowledging its importance to designing learning environments in which it flourishes? The traditional transmission model of teaching continues to dominate many STEM classrooms (Iwuanyanwu, 2020). Knowledge is presented, procedures are demonstrated, and students are expected to absorb, reproduce, and apply. This approach is often defended as efficient. It can cover large amounts of content in limited time. Yet efficiency must not be confused with durability.

Cognitive research repeatedly reminds us that knowledge passively received fades quickly, leaving only fragments where depth once seemed present (Miller, 2003; Spivey, 2023; Yilmaz, 2011). We must therefore ask ourselves: *What remains when the lecture ends?* What becomes the intellectual "gold" that endures beyond assessment cycles? And equally pressing: How do we prevent STEM classrooms from drifting toward intellectual authoritarianism, where knowledge is delivered as settled fact rather than interrogated as reasoned claim? Will we prioritise the delivery of conclusions, or the cultivation of the reasoning that makes conclusions possible? If Socrates were to enter many contemporary STEM classrooms, he might admire the sophistication of content yet question the silence of inquiry. For Socrates, learning began not with telling but with asking. The vitality of inquiry lies in disciplined questioning, probing assumptions, challenging claims, testing coherence. In the natural curiosity of children, he would recognise the seed of epistemic agency. A child asking, "Why?" is already participating in the architecture of argumentation. Our task as STEM educators and practitioners is not to extinguish that instinct, but to discipline it. When we create spaces where students' questions are not interruptions but invitations, argumentation emerges naturally as the medium of thought. The classroom transforms from a site of transmission into a community of inquiry. Here, reasoning and intellect operate together. Students test ideas, justify interpretations, and reflect on alternative explanations. STEM becomes integrative, drawing

Fig. 4.8 Divergent stances on the role of evidence in a STEM task

upon multiple domains while cultivating habits of reasoning, questioning, and reflective judgment. In this light, argumentation becomes STEM Literacy.

This shift is not cosmetic. It repositions STEM literacy as participation in knowledge construction rather than rehearsal of established results. STEM literacy extends far beyond factual recall (Huang et al., 2022). It includes the capacity to evaluate evidence, interrogate claims, and reason through complexity. Yet such capacities do not emerge automatically from exposure to content. They must be intentionally taught, modelled, and scaffolded. To develop argumentation skills, students must learn how evidence functions within disciplines. They must see how data connect to claims, how warrants justify inference, how assumptions shape interpretation, and how contradictions expose fragility in reasoning. These are sophisticated cognitive processes requiring structured practice. When we fail to teach argumentation explicitly, students may rely on intuition, authority, or consensus. When we foreground it, they begin to internalise epistemic norms and recognise that evidence is not decorative; it is foundational. Argumentation skills, therefore, are not auxiliary competencies; they are constitutive of professional formation.

Figure 4.8 introduces us to a collaborative engineering project nearing completion. A design appears promising. Enthusiasm is high. Yet one student, Olisa, raises a concern that the justification lacks credible evidence. The group responds dismissively, “It works. Why do we need more evidence?” The classroom vignette of Olisa crystallises the pedagogical challenge before us. Olisa insists on evidence. His peers resist. This moment is not trivial. It exposes a fault line between epistemic norm and social convenience. We observe two parallel pathways emerging within the same group. One pathway privileges consensus and confidence. The other privileges evidence and coherence. Olisa’s stance aligns with the epistemic commitments articulated in Sect. 4.3, that claims in STEM must be anchored in verifiable grounds. His resistance to peer pressure embodies the intellectual humility and courage necessary for authentic inquiry. If such insistence is experienced as obstruction, we must ask whether argumentation has been taught as a norm or merely mentioned as an expectation.

As educators, we must ask: What culture have we cultivated if students experience requests for evidence as inconvenience? Argumentation must become so deeply embedded in classroom practice that asking for evidence feels natural, not adversarial. This vignette also highlights the social dimension of argumentation. Epistemic courage may require resisting peer pressure. When classrooms normalize critique as constructive rather than confrontational, students like Olisa are protected rather than marginalized. Without explicit scaffolding, through models such as Contiguity Argumentation Theory—CAT (Ogunniyi, 2007) or Toulmin's Argumentation Pattern—TAP (Toulmin, 1958), students often default to intuition or conformity. The question that now remains before us is not whether argumentation matters in STEM. That has been established. The deeper question is whether we are prepared to organise our curricula, our classrooms, and our professional cultures around it.

If we are serious about advancing STEM learning and practice, we must move beyond mere content mastery toward epistemic maturity. The aim is not simply to produce engineers who can calculate, scientists who can model, or technologists who can code. Rather, it is to form professionals who can justify their claims, deliberate responsibly, and decide under conditions of complexity. Accordingly, such professionals demonstrate distinctive intellectual virtues. They welcome critique rather than fear it, recognizing that criticism refines understanding. They revise conclusions when evidence shifts, acknowledging that knowledge is responsive to new data. They navigate obstacles without surrendering integrity, thereby preserving methodological and ethical standards. Finally, they act decisively when warranted, even amid uncertainty, because disciplined reasoning enables responsible judgment. In this sense, argumentation becomes the discipline through which intellect matures into wisdom. Through it, reasoning is tested, evidence is weighed, and judgment is formed. Consequently, STEM education becomes a disciplined, creative, and ethically responsive pursuit of understanding, guided by reason, animated by intellect, and sustained through argumentation.

4.7 Chapter Summary

In this chapter, we traced the epistemic practices of argumentation across reasoning, inquiry, decision-making, and pedagogy in order to clarify its central place within STEM. We began in Sect. 4.2 by distinguishing between choice-making and decision-making. While choices may arise from preference, impulse, or convenience, decisions require justification. They demand the disciplined integration of reasoning and intellect. This foundational distinction established that responsible action in STEM cannot rest on inclination alone; it must be warranted. Section 4.2 therefore laid the conceptual groundwork for understanding why argumentation is indispensable to professional judgment. Our analysis subsequently demonstrated that argumentation is not a peripheral academic exercise or a discrete instructional strategy. It is the organising architecture of STEM thought. From the outset, we recognised reasoning and intellect as interdependent movements within human cognition. Reasoning analyses, differentiates, and advances claims; intellect judges, integrates, and reflects upon their coherence and implications. Their interplay generates insight. Argumentation is the structured dialogue through which this interplay becomes explicit, disciplined, and accountable. Through argumentation, knowledge is not merely received but constructed, tested, contested, refined, and validated.

Across the chapter, a coherent framework emerged. In Sect. 4.3, argumentation was examined as epistemic discipline, the mechanism by which claims become contestable, evidence is scrutinised, and conclusions remain open to revision. In Sect. 4.4, argumentation was explored as disposition, the cultivated posture of curiosity, intellectual humility, and disciplined questioning that characterises the inquiring mind. In Sect. 4.5, we analysed its logical architecture by examining the integration of empirical evidence with conceptual coherence, ensuring that claims are not only supported but rationally warranted. In Sect. 4.6, we turned to pedagogy, the deliberate formation of argumentation as professional habit, requiring explicit modelling, scaffolding, and cultural reinforcement within STEM classrooms. These movements converge into five interrelated dimensions of argumentation: the epistemic dimension, in which argumentation structures knowledge and renders it improvable; the cognitive dimension, through which it refines reasoning, strengthens metacognition, and cultivates intellectual agency; the ethical dimension, by which it grounds responsible decision-making beyond mere preference; the pedagogical dimension, which requires that argumentation be intentionally taught, modelled, and socially protected; and the professional dimension, through which argumentation sustains innovation while maintaining coherence and accountability.

Through this lens, argumentation functions as both epistemic safeguard and practical compass. It enables individuals to interrogate claims, clarify assumptions, apply logic, evaluate evidence, and engage constructively with diverse perspectives. It supports the navigation of uncertainty, the resolution of conflicts of desire or duty, and the transition from analysis to responsible commitment. In doing so, it bridges theory and practice, transforming STEM literacy into participatory competence. The implications for contemporary STEM are profound. We operate in an era of accelerating technological power, artificial intelligence, bioengineering, climate engineering, and autonomous systems. Automation expands capability; it does not guarantee wisdom. The more technologically powerful our tools become, the more indispensable disciplined human judgment becomes. Argumentation ensures that evidence remains central, assumptions are surfaced, decisions are justified, and innovation remains ethically accountable.

4.7.1 Looking Ahead to Chap. 5

This chapter established argumentation as the organising architecture of STEM thought, epistemically structuring knowledge, cognitively refining reasoning, ethically grounding decisions, pedagogically shaping practice, and professionally sustaining innovation. Having clarified why argumentation is indispensable to responsible judgment, we now turn to a pressing contemporary question: *What happens when human reasoning increasingly operates alongside, and sometimes through, machine intelligence?* Chapter 5 extends the argument by situating it within the emerging realities of the Fourth Industrial Revolution. As artificial intelligence, data-driven systems, bioengineering, and autonomous technologies reshape professional and social landscapes, STEM education can no longer focus solely on mastering content or technical procedures. Students must learn to interrogate the assumptions embedded in algorithms, evaluate the evidence behind automated outputs, examine the values coded into technological systems, and take intellectual responsibility in environments where decisions are increasingly mediated by machines.

If this chapter established that argumentation transforms choice into justified decision, Chap. 5 examines how this transformation becomes even more urgent in contexts shaped by the mind–machine nexus. When automation expands capability without guaranteeing wisdom, disciplined human judgment becomes the stabilising force. Argumentation functions here not only as epistemic safeguard but as transformative praxis. It equips learners to critique machine-generated claims, question data integrity, surface hidden biases, and navigate complex socio-technical systems with ethical clarity.

Chapter 5 therefore moves from the architecture of argumentation to its application in technologically saturated environments. It situates argumentation within design thinking, ethical reasoning, and technological literacy, demonstrating that future-ready STEM professionals must be capable of engaging both human and machine forms of intelligence. In this way, Chap. 5 deepens the claim advanced in this chapter that argumentation is not merely foundational to STEM knowledge; it is indispensable to shaping reflective, ethical, and epistemically responsible professionals in an age of accelerating technological power.

4.8 Reflection Questions

The following questions are intended to prompt intellectual consolidation and professional self-examination.

Conceptual Foundations (Sect. 4.2)
1. What distinguishes a *choice* from a *decision* in professional STEM contexts?
2. In what ways does justification transform preference into responsible judgment?

3. How do reasoning and intellect function differently yet interdependently in your own decision-making?

Epistemic Dimension (Sect. 4.3)

4. How does argumentation render knowledge contestable and improvable rather than fixed?
5. What risks emerge when claims are accepted without structured scrutiny?
6. How can you create environments where evidence is expected rather than optional?

Dispositional Dimension (Sect. 4.4)

7. Which intellectual virtues (e.g., humility, perseverance, curiosity) are most essential for sustained inquiry in STEM?
8. How can argumentation be cultivated as a disposition rather than performed as a task?

Logical Architecture (Sect. 4.5)

9. What is the relationship between empirical evidence and conceptual coherence in constructing a strong argument?
10. How do hidden assumptions weaken arguments, even when data appears robust?
11. In your discipline, what counts as credible evidence—and why?

Pedagogical Dimension (Sect. 4.6)

12. How can teachers model argumentation as professional habit rather than classroom ritual?
13. What cultural norms must be established to protect students who insist on evidence, like Olisa?
14. How might assessment practices either reinforce or undermine epistemic responsibility?

Ethical and Professional Implications

15. Why is argumentation essential for innovation under conditions of uncertainty?
16. How does argumentation function as both safeguard and compass in contemporary STEM practice?
17. In what ways does your current professional preparation emphasise technical proficiency over epistemic responsibility, and how might this be recalibrated?

Integrative Reflection

18. How has your understanding of STEM shifted, from mastering content to participating in a disciplined practice of justified reasoning?
19. What concrete steps can you take to strengthen argumentation as a cognitive, ethical, and professional commitment?
20. How might the absence of argumentation compromise not only knowledge production but societal trust in STEM?

References

Abney, D. H., Dale, R., Yoshimi, J., Kello, C. T., Tylén, K., & Fusaroli, R. (2014). Joint perceptual decision-making: A case study in explanatory pluralism. *Frontiers in Psychology, 5*, 330.

Aslam, S., Alghamdi, A. A., Abid, N., & Kumar, T. (2023). Challenges in implementing STEM education: Insights from novice STEM teachers in developing countries. *Sustainability, 15*(19), 14455. https://doi.org/10.3390/su151914455

Dönmez, İ., Gülen, S., & Ayaz, M. (2022). Impact of argumentation-based STEM activities on ongoing STEM motivation. *Journal for STEM Education Research, 5*, 78–101. https://doi.org/10.1007/s41979-021-00062-2

Einstein, A. (1920). *Relativity: The special and general theory* (R. W. Lawson, Trans.). Henry Holt and Company.

Erduran, S., & Park, W. (2023). Argumentation in physics education research: Recent trends and key themes. In M. F. Taşar & P. R. L. Heron (Eds.), *The international handbook of physics education research: Learning physics* (pp. 16–32). AIP Publishing.

Huang, X., Erduran, S., Luo, K., Zhang, P., & Zheng, M. (2022). Investigating in-service teachers' STEM literacy: The role of subject background and gender. *Research in Science & Technological Education, 42*(3), 867–887. https://doi.org/10.1080/02635143.2022.2153243

Iwuanyanwu, P. N. (2017). *An analysis of pre-service teachers' ability to use a dialogical argumentation instructional model to solve mathematical problems in physics*. University of the Western Cape. (Thesis-PhD). https://etd.uwc.a.za/handle/11394/6252

Iwuanyanwu, P. N. (2020). Nature of problem-solving skills for 21st century STEM learners: What teachers need to know. *Journal of STEM Teacher Education, 55*(1), 27–40. https://doi.org/10.30707/JSTE55.1/MMDZ8325

Iwuanyanwu, P. N. (2022). What students gain by learning through argumentation. *International Journal of Teaching and Learning in Higher Education, 34*(1), 97–107.

Iwuanyanwu, P. N. (2023). When science is taught this way, students become critical friends: Setting the stage for student teachers. *Research in Science Education, 53*(6), 1063–1079. https://doi.org/10.1007/s11165-023-10122-9

Koichu, B., Schwarz, B. B., Heyd-Metzuyanim, E., Tabach, M., & Yarden, A. (2022). Design practices and principles for promoting dialogic argumentation via interdisciplinarity. *Learning, Culture and Social Interaction, 37*, 100657. https://doi.org/10.1016/j.lcsi.2022.100657

Miller, G. A. (2003). The cognitive revolution: A historical perspective. *Trends in Cognitive Sciences, 7*(3), 141–144.

Mahmood, A., Huang, X., & Rehman, N. (2025). STEM education as a catalyst for career aspirations and 21st-century competences: Insights from teachers' perspectives. *School Science and Mathematics*, 1–15. https://doi.org/10.1111/ssm.18381

National Research Council. (2011). *Successful K-12 STEM education: Identifying effective approaches in science, technology, engineering and mathematics*. National Academies Press.

Ogunniyi, M. B. (2007). Teachers' stances and practical arguments regarding a science-indigenous knowledge curriculum: Part 1. *International Journal of Science Education, 29*(8), 963–986. https://doi.org/10.1080/09500690600931020

Ogunniyi, M., & Iwuanyanwu, P. N. (2024). Analysis of teachers' perspectives towards the use of IKS to improve STEM education for sustainable development. *African Journal of Research in Mathematics, Science and Technology Education, 28*(3), 319–329. https://doi.org/10.1080/18117295.2024.2352980

Osborne, J. F. (2019). Not "hands on" but "minds on": A response to Furtak and Penuel. *Science Education, 103*(5), 1280–1283. https://doi.org/10.1002/sce.21543

Spivey, M. J. (2023). Cognitive science progresses toward interactive frameworks. *Topics in Cognitive Science, 15*(2), 219–254. https://doi.org/10.1111/tops.12645

Tarekegn, G., Osborne, J., & Tadesse, M. (2022). Impact of dialogic argumentation pedagogy on grade 8 students' epistemic knowledge of science. In M. Kalogiannakis & M. Ampartzaki

(Eds.), *Advances in research in STEM education*. IntechOpen. https://doi.org/10.5772/intechopen.104536

Toulmin, S. E. (1958). The uses of argument. Cambridge University Press.

Toulmin, S. E., Rieke, R. D., & Janik, A. (1979). An introduction to reasoning. Macmillan.

Tytler, R. (2020). STEM education for the twenty-first century. In J. Anderson & Y. Li (Eds.), *Integrated approaches to STEM education: An international perspective* (pp. 21–43). Springer International. https://doi.org/10.1007/978-3-030-52229-2_3

Wright, K. L., Wiggins, L., & Miller, S. (2025). Becoming STEM writers: A low-cost intervention for teaching argumentation. *Journal of College Science Teaching, 54*(6), 519–524. https://doi.org/10.1080/0047231X.2025.2507003

Yilmaz, K. (2011). The cognitive perspective on learning: Its theoretical underpinnings and implications for classroom practices. *The Clearing House, 84*(5), 204–212. https://doi.org/10.1080/00098655.2011.568989COGENTEDUCATION17

Chapter 5
The Mind–Machine Nexus and the Praxis of Transformative Argumentation in STEM Education

5.1 Introduction

In this chapter, we extend our inquiry into the epistemic foundations of STEM by examining the role of argumentation in the knowledge practices that define contemporary scientific and technological life. Building on the earlier chapters, we can begin to see that argumentation is not merely an instructional technique or a communication skill but an essential epistemic engine through which STEM practitioners, and the students who aspire to join them, make sense of the world. Argumentation cultivates the habits of mind that the twenty-first century demands. This includes the ability to reason critically, evaluate evidence, articulate claims, and engage constructively with competing interpretations. These are not optional capacities. They are the preconditions for participating responsibly in a world increasingly shaped by frontier technologies.

As we reflect together on the pedagogical and philosophical demands of STEM education today, we recognize that argumentation is indispensable for preparing students to navigate the epistemic complexities of the Fourth Industrial Revolution (4IR). The 4IR is defined not only by digital transformation but by the merging of human cognition with machine intelligence, the convergence of biological and computational systems, and the pervasive automation of decision-making processes. In such a landscape, students must learn not only what technologies do but how to interrogate the assumptions, values, and evidence that underpin them. The capacity to construct well-reasoned arguments, challenge data patterns generated by AI, scrutinize the ethical implications of gene editing, and evaluate the reliability of sensor-generated data in IoT systems becomes central to a human-centered technological literacy.

Argumentation also offers students a space to articulate their own ideas and to grapple with the conceptual, ethical, and methodological tensions that arise in STEM. When students pose questions, challenge established notions, or propose

P. N. Iwuanyanwu, *Empowering STEM Thinkers Through Argumentation*, Contemporary Trends and Issues in Science Education 69,
https://doi.org/10.1007/978-3-032-27906-4_5

explanations rooted in evidence, they participate in the same epistemic practices that animate scientific inquiry itself. Through this process, they not only consolidate their evolving understanding but also learn to communicate it to peers, educators, and the broader public. In this way, argumentation becomes a bridge between disciplinary knowledge and societal relevance, a bridge that grows ever more vital as scientific knowledge becomes entangled with political, economic, and ethical debates surrounding frontier technologies. Moreover, as we consider the broader implications, it becomes clear that emphasizing argumentation in STEM education extends well beyond classroom practice.

A society capable of engaging critically with emerging technologies, questioning algorithmic decisions, evaluating genetic risks, or debating the merits of automation, depends on citizens who can reason with evidence and articulate well-founded judgments. When students learn to argue well, they learn to think well; and when they learn to think well, they are better prepared to act responsibly in a world where technological innovation often outpaces regulatory, moral, and even cultural reflection. Thus, in situating argumentation at the heart of STEM education, we are invited to embrace not a rhetorical add-on but a fundamental epistemic orientation, one that affirms the centrality of human judgment in an era increasingly shaped by non-human intelligence. Argumentation becomes the means through which we resist the temptation to surrender our reasoning to algorithms and the means through which we assert the moral and intellectual agency that must remain at the center of a human-centered technological future.

5.2 Fundamental Understanding of Argumentation Framework

To develop a rigorous understanding of argumentation within STEM, we first distinguish clearly between argument and argumentation. An argument refers to the act of presenting reasons either in support of or against a proposition or course of action (Britannica, 2008; Toulmin, 2003). This tradition has deep philosophical roots. Thinkers such as Socrates, Plato, and Aristotle, writing between 470 and 320 B.C., laid the earliest foundations for analysing arguments as structured, rational acts of inquiry. Aristotle's contribution remains especially instructive: his tripartite model of minor premise, major premise, and conclusion represents one of the earliest attempts to capture the microstructure of reasoning. It also foregrounds an epistemic point that is as vital today as it was in antiquity, the truth-value of premises determines the validity of the conclusion. A true premise may anchor a sound argument; a false one may render even the most elegant conclusion fallacious.

Yet as we reflect on the complexity of contemporary STEM reasoning, particularly in an age of frontier technologies such as AI, CRISPR gene editing, robotics, and the Internet of Things, we find ourselves questioning whether such a classical structure is sufficient. The simplicity of Aristotle's schema is philosophically

Table 5.1 Description of the elements of argument

Elements of arguments	Description of elements of argument
Claim	A claim is an assertion or statement that is being argued for or against. It is the main point or position that the argument is trying to prove or support. However, the merit of the claim is often in question and relies heavily on the availability of evidential support.
Data	Data refers to the evidence or information that is used to substantiate the claim. It can include facts, statistics, research findings, expert opinions, or any other relevant information that supports the argument.
Warrant	A warrant is a statement or set of statements that acts as the bridge between the evidence and the claim. It ensures that there is a logical and coherent connection between the two. The warrant explains why the evidence is relevant and how it supports the claim.
Backing	Backing consists of statements that support the validity of the warrants. It reinforces the overall strength of the argument by providing additional evidence or reasoning that supports the warrants.
Qualifier	Qualifier includes statements that define the conditions under which the claim remains valid. It acknowledges any limitations or exceptions that may exist and provides context for the claim. Qualifiers help to clarify the scope and applicability of the claim.
Rebuttal	Rebuttal consists of statements that challenge or refute the claim. It highlights potential weaknesses or flaws in the argument and presents counterarguments or alternative perspectives. Rebuttals are important for considering opposing viewpoints and strengthening the overall argument.

elegant, but our epistemic landscape is no longer simple. Algorithms now generate premises from data sets whose provenance is often opaque. Genetic engineering technologies force us to evaluate claims whose ethical and biological implications extend beyond traditional logical form. Under such conditions, the classical triptych of minor premise, major premise, and conclusion becomes a starting point rather than a satisfactory endpoint.

It is precisely this epistemic pressure that Stephen Toulmin sought to address through his Toulmin Argument Pattern (TAP). Toulmin's model offers a more elaborate framework for understanding arguments, recognizing that real-world reasoning rarely conforms to perfectly formed syllogisms. By foregrounding claims, data, warrants, backings, qualifiers, and rebuttals, TAP acknowledges the complexity, uncertainty, and social embeddedness of argumentation as demonstrated in Table 5.1. It also underscores an essential distinction. While arguments involve the expression of reasons, argumentation is the process of assembling, justifying, and evaluating those reasons through evidence (Kuhn & Udell, 2007). In this sense, argumentation is epistemic work that becomes increasingly urgent when STEM claims intersect with data-driven machine predictions, bioethical dilemmas, or automated decision-making systems.

Although TAP remains influential, its limitations must also be acknowledged. One of its philosophical strengths is the focus on the assembly of argumentative components and the source of epistemic questions. Toulmin's model tends to assume a passive opponent, an assumption that simplifies the interactive nature of discourse. Yet research by Erduran et al. (2004) reminds us that argumentation is not a unidirectional act but a dialogic one. When two individuals debate the safety of CRISPR, or the reliability of an AI's predictive model, they do not merely present claims, they interrogate, resist, refine, and reinterpret one another's warrants and data. Ignoring this active engagement risks flattening the dynamics of real argumentative exchange.

Further, scholars have identified conceptual overlaps within TAP, particularly between the meanings of claim and warrant, that may obscure rather than clarify the internal logic of complex arguments (Van Eemeren et al., 1987). Such ambiguities become more pronounced when arguments concern emerging technological contexts where the boundary between data and warrant is increasingly difficult to discern. For instance, when AI models generate patterns without transparent justification, is the model's output "data," or does it operate as an implicit "warrant" grounded in statistical inference? Such questions challenge the adequacy of TAP as a universal model.

Another important limitation concerns TAP's cultural and contextual narrowness. Toulmin's model privileges forms of reasoning that align with logical structures, often sidelining other forms of argumentation that are culturally embedded, informal, value-laden, or narrative in nature (Ogunniyi, 2007a). Yet in many socio-STEM contexts, such as community debates on gene editing, environmental justice, or AI surveillance, arguments are not confined to formal logic. They are infused with ethical convictions, cultural presuppositions, and lived experience. Ogunniyi (2007a, b) rightly critiques TAP for neglecting these dimensions, arguing that such neglect distorts our understanding of how people actually reason about scientific issues, particularly in cross-cultural settings.

To address these gaps, Ogunniyi introduced the Contiguity Argumentation Theory (CAT), which identifies three major categories of arguments: logical, non-logical, and value-laden (Ogunniyi, 2007a, b). CAT acknowledges that argumentation is not merely a logical activity but a cognitive and cultural one. Its framework allows us to examine how individuals shift between these categories, sometimes privileging empirical evidence, sometimes moral commitments, sometimes cultural narratives, especially in debates involving socio-scientific or frontier-technology issues. For instance, a discussion about CRISPR gene editing will naturally evoke not only logical arguments about biological risks but value-laden arguments about ethics, identity, and human dignity. Similarly, debates about AI autonomy can invoke non-logical concerns grounded in fear, trust, or social narratives.

By recognising these shifts, CAT provides a more expansive and human-centered account of argumentation that is essential for navigating the complex, ethically fraught terrain of modern STEM fields. It invites us to examine argumentation not merely as a structure but as a lived epistemic practice shaped by logic, culture, emotion, and moral reasoning. In an age defined by frontier technologies, this broader framework is indispensable, for it allows us to understand how people actually reason in contexts where evidence, values, and technology intersect.

5.3 The Contiguity Argumentation Theory

To deepen our inquiry into argumentation as an epistemic practice, we now turn to the Contiguity Argumentation Theory (CAT), a framework whose philosophical roots trace back to the classical teachings of Plato and Aristotle. The Aristotelian notion of contiguity holds that certain mental states, ideas, or systems of thought have a natural tendency to evoke one another, producing heightened awareness and new cognitive associations (Ogunniyi, 2007a, b). This principle-association through conceptual proximity, sets the stage for understanding how ideas shift, transform, or destabilize in the course of argumentation. Whereas Toulmin's Argument Pattern foregrounds the logical anatomy of an argument, the Contiguity Argumentation Theory extends our analytical reach. CAT acknowledges what Toulmin leaves largely unexamined. This include the non-logical, metaphysical, axiological, and emotional dimensions that shape how individuals reason, particularly in dialogic and culturally diverse STEM environments. As we navigate a world defined by frontier technologies, we encounter AI capable of generating probabilistic predictions, CRISPR reshaping the boundaries of biological possibility, IoT systems producing continuous data streams, and robotics transforming human labour. In this context, we cannot ignore the social, ethical, and affective forces that structure how people argue about scientific and technological issues. CAT invites us to account for these dimensions.

By engaging with these broader layers of meaning, CAT offers a more comprehensive understanding of argumentation and the cognitive shifts individuals undergo when contexts change or when they encounter new or disruptive perspectives. Ogunniyi (2007a, b) identifies five worldview elements that shape argumentation dynamics. This includes: the Dominant worldview, which governs the prevailing perspective in a discourse; the Suppressed worldview, which is subordinated by the dominant frame; the Assimilated worldview, which concedes to the dominant perspective; the Emergent worldview, which arises from new cognitive or experiential encounters; and the Equipollent worldview, in which two competing perspectives exert equal cognitive force. These interwoven worldviews illustrate how the use of argumentation in STEM education is not merely a logical act but a dynamic negotiation of identity, belief, evidence, and emotional salience (Iwuanyanwu, 2020).

This worldview framework becomes particularly important when we consider the realities of twenty-first-century STEM education. Despite significant attention to STEM pedagogy, we still know too little about how argumentation skills prepare students for their lives beyond graduation, lives that will be shaped by ubiquitous automation, algorithmic decision-making, and perpetual technological disruption. Similarly, we are only beginning to understand how prepared STEM teachers are to cultivate argumentation skills that enable students to engage critically with the socio-technical challenges brought about by artificial intelligence, genetic engineering, climate technologies, and other frontier innovations. This gap matters. If educators are to prepare students (future STEM professionals) for a human-centered technological future, they must recognize that argumentation is not simply a method for academic reasoning; it is a cognitive tool for navigating uncertainty, ethical complexity, and contested evidence. When students reason about the risks of autonomous weapons, the social implications of AI-generated misinformation, or the ethics of embryo editing with CRISPR, they must be able to shift among logical, non-logical, and value-laden arguments. CAT provides a conceptual lens for analysing these shifts. A fuller discussion of how CAT and TAP operate within classroom settings will follow later in this chapter.

From these considerations emerges a guiding principle, any argumentation framework must be no less complex than the arguments it seeks to analyse. Arguments are fundamentally subjective engagements with objective or semi-objective situations. Individuals may share the same goals yet arrive at radically different interpretations because perception itself is fluid, contextually shaped, and emotionally charged. As highlighted in Chap. 3, whether through Jason's eyewitness account of the confrontation between Chris and David or the contrasting interpretations of Steve and Sam regarding the English Premier League, arguments are rarely reducible to purely logical form. They are always animated by perception, emotion, value commitments, and social identity.

A particularly vivid illustration of this principle is depicted in Fig. 5.1. The figure presents the anecdote of the man who deliberately painted one side of his car pink and the other dark blue in order to stage a scenario where eyewitnesses would inevitably contradict one another following an accident. His deliberate design exposes how perceptions, far from being uniform or neutral, are shaped by selective cognitive attention, emotional states, and positionality. One witness may confidently report that the car was pink; another may be equally insistent that it was blue. Both believe they "see the truth," and indeed both do, but only partially. Each witness sees one side of a larger reality, and their respective claims, though contradictory,

Fig. 5.1 Anecdote of controversial situation

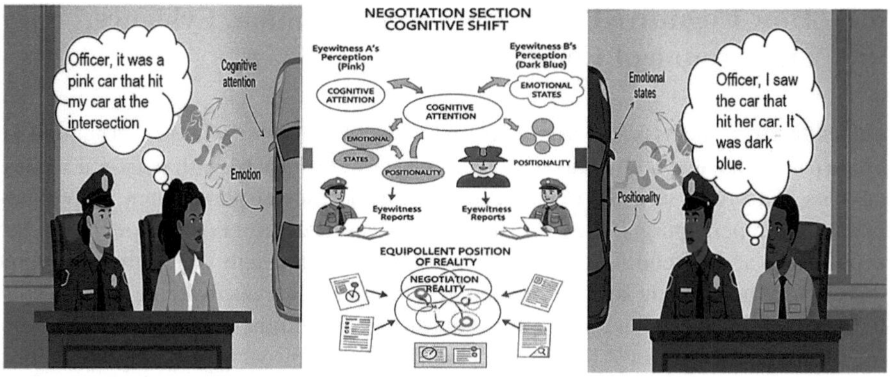

Fig. 5.2 Eyewitness reporting and cross-examination

emerge from their own valid but incomplete cognitive positions. What follows is a clash of logical, non-logical, and value-laden arguments, all originating from the same physical event. This example underscores why CAT is an essential analytical framework. It helps to illuminate the cognitive shifts individuals make as they interpret, justify, and defend their perspectives in dialogic exchange.

In light of the foregoing discussion, the cognitive shifts made by the eyewitnesses as they debate the colour of the car provide a powerful illustration of how human reasoning operates across logical, non-logical, and affective dimensions. Each eyewitness forms a perspective grounded not only in perception but also in memory, emotion, and implicit beliefs about what counts as credible evidence. Through the lens of CAT, we can trace how individuals navigate between worldviews as they articulate claims and defend them under scrutiny. This process is visually elaborated in Fig. 5.2, which captures the moment when the witnesses present their accounts to law enforcement officers. During this cross-examination phase, the first eyewitness, a lady who observed only the pink side of the vehicle, clings firmly to her dominant perspective, asserting, "Officer, it was a pink car that hit my car." On the other hand, the second eyewitness, who observed the dark blue side, confidently reports, "Officer, it was a dark blue car that hit her car." In both cases, the eyewitnesses' perceptual frames override any alternative interpretation. Thus, perception is effectively suppressed in favour of what each believes is a socially dominant or personally salient narrative of the event.

Here, the officers must scrutinise the Data (the eyewitness reports) and negotiate an Equipollent position, one in which both interpretations remain cognitively possible and neither is prematurely privileged. This negotiation process, as CAT demonstrates, is deeply epistemic. It reveals how individuals move across cognitive and affective states as they attempt to reconcile competing truths. The simple anecdote of a two-coloured car thus becomes emblematic of a broader truth, namely, that in a world shaped increasingly by frontier technologies, our capacity to argue critically, reflectively, and ethically may be among the most important skills we can cultivate.

5.4 How Cognitive Bias Shapes Interpretation, Evidence, and Argument Construction

Understanding the type of argument each witness offers, whether grounded in empirical logic, shaped by narrative interpretation, or infused with personal values, enables us to recognise the deeper reasoning processes at play. When individuals debate their perceptions, they are not merely disagreeing about colour; they are articulating and defending the cognitive frameworks through which they interpret the world. These subtle shifts mirror the reasoning challenges that characterise contemporary STEM debates, from ethical questions concerning CRISPR genome editing to the reliability of AI-driven decision-making. In each case, argumentation becomes a window into how people negotiate uncertainty, reconcile competing ideas or worldviews, and determine what counts as sufficient justification. Regardless of the position an eyewitness ultimately defends, the dialogic process itself is epistemically productive. Through argumentation, individuals test the coherence of their claims, evaluate alternative interpretations, and seek validation from an audience. In doing so, they enact precisely the kind of cognitive work required in STEM disciplines, scrutinising evidence, reflecting on assumptions, and refining claims in response to challenge. This recursive cycle of claim, critique, and revision constitutes the engine that drives scientific inquiry, technological innovation, and ethical deliberation.

However, the complexity of argumentation becomes even more apparent when we consider the levels of argumentative sophistication demonstrated in STEM classrooms. Research by Erduran et al. (2004) and Simon and Johnson (2008) shows that arguments vary widely, from simple assertions to increasingly complex structures involving rebuttals and counter-rebuttals. Table 5.2 shows how argumentative complexity unfolds across five levels, from non-oppositional arguments to arguments containing multiple strong rebuttals. Developing an understanding of these levels is crucial for cultivating students' STEM reasoning. When students differentiate between non-oppositional arguments and those with valid grounds, they begin to appreciate the role of evidence and logical coherence. As they encounter weak and strong rebuttals, they learn to anticipate criticism, strengthen their claims, and refine their reasoning processes. When they confront multiple strong rebuttals,

Table 5.2 Levels of argumentation pattern

Levels	Description of argumentation
1	Non-oppositional arguments with simple claims or counterclaims.
2	Arguments with claims supported with valid grounds (data, warrants, backings and qualifiers) but with no rebuttals.
3	Arguments with claims supported with valid grounds and only occasional weak rebuttals.
4	Arguments with claims supported with valid grounds and at least one strong rebuttal.
5	Arguments with claims supported with valid grounds and with two or more strong rebuttals.

they are challenged to move beyond binary thinking, synthesizing diverse perspectives and engaging in rigorous dialectical reasoning.

These abilities are not merely academic. They foster a culture of critical friendship, reflective inquiry, and collaborative knowledge-building within STEM communities. As Iwuanyanwu (2023) argues, such capacities empower students to engage constructively with their peers, challenge assumptions responsibly, and contribute meaningfully to disciplinary and societal knowledge production. In an era where frontier technologies continually redraw the boundaries of what is possible, argumentation becomes not just a skill but a form of epistemic resilience, an essential tool for engaging thoughtfully, ethically, and creatively with an evolving technological world. It is through argumentation that students learn how to articulate their positions, encounter alternative viewpoints, and reassess their initial assumptions. More importantly, argumentation deepens understanding. When individuals enter into structured discourse, they do more than defend their own viewpoints, they encounter the limits of their perspectives and expand their conceptual repertoire. In STEM contexts, where students must navigate complex, ambiguous, and ethically charged issues, such as AI surveillance, algorithmic bias, autonomous robotics, or the unintended consequences of IoT interconnectivity, levels of argumentation pattern (Erduran et al., 2004) enable them to reason through these tensions with intellectual humility and epistemic responsibility.

Furthermore, non-oppositional arguments represent the most foundational form. Here, students offer evidence and reasoning to defend a claim without directly countering any opposing perspective. Such arguments teach students how to articulate STEM ideas clearly and persuasively, without immediately defaulting to adversarial stances. This is particularly important in interdisciplinary fields shaped by frontier technologies, AI ethics, genetic engineering, autonomous robotics, where collaborative sense-making is as valuable as debate. As students progress, they encounter arguments with valid grounds, in which claims are supported by solid evidence and logical coherence. Mastery of this level signals an emerging capacity to distinguish between well-founded and loosely constructed assertions, an essential skill at a time when misinformation, algorithmic opacity, and techno-solutionism increasingly challenge public STEM literacy.

The next level involves arguments with valid grounds and weak rebuttals. Students begin to acknowledge counterarguments but do so tentatively or superficially. This stage is important because it reveals a developing awareness of contestation, even if the responses lack epistemic force. In STEM contexts, where competing interpretations of data, model uncertainty, or design trade-offs are unavoidable, students should learn not merely to recognise counterarguments but to treat them as opportunities for strengthening their reasoning. With valid grounds and strong rebuttals, students encounter a more demanding form of argumentative engagement. Here, they should confront compelling counterarguments that meaningfully challenge their initial claims. This level cultivates intellectual resilience: students learn to refine their reasoning, anticipate objections, and marshal stronger evidence. These capacities mirror the iterative processes of scientific inquiry, engineering design, and ethical deliberation in frontier technology domains, where claims must continually withstand scrutiny.

Finally, arguments with valid grounds and multiple strong rebuttals represent the apex of argumentative complexity. Students should weigh multiple competing perspectives, evaluate their strength, and justify their stance in relation to a constellation of alternatives. This level mirrors the real-world decision-making contexts found in advanced STEM fields, balancing the risks and promises of AI deployment, negotiating ethical boundaries of CRISPR applications, or assessing tradeoffs in large-scale IoT infrastructures. Engaging at this level requires epistemic humility, analytical rigour, and an appreciation for pluralism in evidence and values. As students move across these levels, they not only deepen their conceptual understanding but also develop the intellectual dispositions that sustain a culture of critical inquiry within STEM education. They learn to provide constructive feedback, challenge ideas respectfully, and articulate claims with an awareness of nuance and complexity. This growth reflects what Iwuanyanwu (2023) identifies as the development of agency and critical friendship, dispositions that empower students to participate in scientific and technological discourse with confidence and ethical awareness. Ultimately, cultivating these layered argumentative capacities equips students with a crucial set of epistemic tools. It enables them to engage thoughtfully with the complexities of frontier technologies, confront ambiguous evidence responsibly, and contribute to the advancement of STEM knowledge in ways that are both intellectually rigorous and socially conscientious.

If STEM education is to prepare professionals for the complexities of contemporary technological life, then argumentation must be understood as one of its most powerful pedagogical affordances. Far from being a peripheral skill, argumentation operates as an epistemic, dialogic, and ethical practice through which students learn to reason carefully, interrogate evidence, and situate knowledge claims within broader social and technological contexts. This is particularly urgent in an era shaped by frontier technologies, artificial intelligence, genetic engineering such as CRISPR, the Internet of Things, and autonomous robotics, each of which complicates traditional notions of truth, evidence, and responsibility. As research has consistently shown (e.g., Asterhan & Schwarz, 2007; Erduran & Park, 2023; Iwuanyanwu & Ogunniyi, 2020; Lubben et al., 2010; Ogunniyi, 2007a, b; Osborne, 2019; Schwarz et al., 2017), argumentation cultivates habits of mind that lie at the heart of intellectual autonomy. When individuals engage in the discursive practices of making claims, examining evidence, challenging assumptions, and responding to counterarguments, they participate in the same forms of inquiry that structure scientific and technological advancement. Yet the value of argumentation extends beyond cognitive development; it also enables learners to confront the ethical and epistemic tensions inherent in contemporary STEM landscapes.

Consider the dilemmas introduced by AI-driven decision-making systems. These tools promise efficiency and predictive accuracy, yet students should learn to question what counts as "evidence" in a machine-learning model, how algorithmic bias emerges, and what ethical frameworks should govern automated decisions that affect human lives. Argumentation becomes the intellectual space in which such questions can be explored critically rather than taken for granted as inevitable features of technological progress. Similarly, in the age of CRISPR and other gene-editing technologies, students must grapple with questions that are both scientific and moral: What constitutes a legitimate use of gene editing? How do we evaluate risks that extend across generations? Who has the authority to make decisions on behalf of future life? These are questions with no simple answers; they require the deliberative, reflective, and evaluative capacities that argumentation uniquely cultivates.

The same is true of the vast, sensor-saturated environments created by the Internet of Things. With data flowing continuously between devices, infrastructures, and human bodies, students should learn to reason about accuracy, privacy, security, and the interpretation of complex datasets. In such settings, argumentation becomes a tool for managing epistemic uncertainty and making judicious decisions in contexts where information is abundant but not always reliable. Robotics, too, raises deeply philosophical questions as autonomous systems enter workplaces, homes, and public spaces. Students should learn to weigh competing claims about safety, labour displacement, human–machine collaboration, and the moral status of autonomous agents. Argumentation provides the discursive structure through which such evaluations can occur responsibly and transparently.

In the STEM classroom, therefore, argumentation enables far more than the exchange of ideas; it cultivates a disciplined form of intellectual engagement that allows learners to question, interpret, and construct knowledge within technologically complex environments. It invites students to articulate their assumptions, assess the credibility of evidence, and refine their views in response to reasoned critique. It also positions teachers as epistemic facilitators who guide learners toward deeper and more coherent forms of understanding, rather than as transmitters of predetermined content. Most importantly, argumentation creates a learning environment defined by inquiry rather than passive consumption. Students begin to see their own reasoning as central to meaning-making, rather than as an optional supplement to technical knowledge. They learn to collaborate dialogically, to listen for nuance, to negotiate disagreement, and to recognise the legitimacy of alternative perspectives. As Iwuanyanwu (2023) argues, these practices foster dispositions of critical friendship, intellectual humility, and reflective judgment, qualities without which STEM education risks devolving into rote proceduralism or uncritical technological optimism.

In this way, argumentation serves as both a cognitive and moral apprenticeship for the age of frontier technologies. It shapes individuals who can think deeply, question courageously, deliberate ethically, and participate responsibly in scientific and technological communities. It is through argumentation that students come to appreciate STEM not merely as a collection of facts or tools but as a profoundly human endeavour grounded in reasoning, evidence, creativity, and ethical reflection.

5.5 Argumentation as a Framework for Engineering Design

Argumentation becomes most powerful when it is not treated merely as an abstract epistemic exercise but as a living method through which ideas evolve, designs take shape, and engineering problems are solved with integrity and imagination. The engineering design task presented here, a gravity-powered cargo vehicle capable of navigating slopes, responding to obstacles, and launching its payload toward a target, offers us a practical space in which to demonstrate how argumentation functions as a cognitive engine for STEM learning. We view design not simply as the construction of artefacts but as a philosophical activity, one that calls upon individuals to reason, imagine, deliberate, and justify. To illustrate how argumentation permeates engineering practice, we structure this section into two parts: (1) the design problem as a site of epistemic reasoning, and (2) argumentation across the phases of engineering work. Together, these sections show how imagination, design, testing, revision, and presentation form a continuous argumentative cycle, one that refines both knowledge and judgment.

5.5.1 The Engineering Problem as a Site of Epistemic Reasoning

Engineering design provides a fertile space for illustrating how argumentation operates as an epistemic, creative, and ethical practice. When we ask students, or professional teams, to design a gravity-powered cargo vehicle capable of descending a sloped track, navigating obstacles, and launching a payload toward a designated target zone, we are doing far more than assigning a technical exercise. We are calling them into a mode of reasoning where claims must be justified, decisions defended, uncertainties interrogated, and ethical implications made explicit. The design task becomes a microcosm of STEM inquiry, one in which argumentation is not an add-on but the very grammar through which design thinking unfolds. The design challenge begins with a transdisciplinary problem:

> *Design a gravity-powered cargo vehicle that travels down a sloped track, overcomes or navigates around obstacles, and, when its path is blocked, launches its cargo accurately toward a target zone.*

The problem immediately invokes a constellation of STEM concepts:

(i) gravitational acceleration and energy transfer,
(ii) mechanical stability and structural integrity,
(iii) friction and surface interaction,

(iv) projectile motion and force transmission,
(v) geometric reasoning, model construction, and algorithmic planning.

But beyond disciplinary content, the task also opens a space for epistemic activity. Students or a team of professionals must justify why a particular ramp angle is optimal, why one wheel arrangement offers greater stability than another, or why a lever-based launching mechanism is preferable to a spring-based system. These justifications require claims, supported by evidence, connected by reasoning, and tested through counterclaims and rebuttals.

To begin, we invite the team to imagine the problem space, sketch the possible obstacles, and visualise how the vehicle interacts with its environment. This imaginative phase is not cognitively trivial; it mirrors the abductive reasoning that underlies scientific discovery. It is here that argumentation first emerges in intra-locutory form, silent reasoning, internal deliberations, private mental simulations. A team working on the project may ask themselves: *What if the slope is too steep? What if frictional forces destabilise the cargo? What if the obstacle blocks both the direct and alternate paths?* At this point, imagination becomes the seedbed for argumentation.

Before any building begins, teams must imagine the scenario. They sketch slopes, obstacles, wheel configurations, and launching mechanisms. They speculate about failure modes, potential hazards, and ways to improve stability. Figure 5.3 illustrates the imagination phase, where teams sketch possible simulated vehicle along with the cargo, obstacles, and potential launch scenarios, and visualize how the prototype might operate. These early sketches serve not merely as drawings but as philosophical propositions, hypotheses about how the world might behave under certain constraints. We reference them deliberately to underscore that imagination is not mere creativity,

Fig. 5.3 Visualization of the Simulated Vehicle along with the Cargo, Obstacle, and Possible Launch Scenario

it is a structured form of anticipatory reasoning. In this sense, asking critical questions and seeking their answers are paramount. The imaginative phase and its anticipatory reasoning necessitate questions such as: (1) What if the ramp is too steep? (2) What if the wheels lose traction? (3) What if the cargo shifts its center of mass? At this stage, argumentation is largely intra-locutory. The team debates silently within themselves, weighing alternatives, anticipating failures, and narrating reasons that may later be shared publicly. This form of individual reasoning mirrors the imaginative processes used by scientists and engineers when first modelling a system or drafting an idea.

Ethical reasoning in the design space is highly significant. Beyond physics and mechanics, the team must implicitly confront ethical considerations. Even a cardboard prototype raises questions that resonate with frontier technology domains: (1) Should we prioritize accuracy or safety? (2) How do we balance efficiency with reliability? (3) What risks are acceptable? Who decides? Such questions mirror real engineering dilemmas in autonomous vehicles, drone delivery systems, gene editing protocols, and large-scale IoT networks. We cultivate the habit of ethical reflection not through lectures but through design contexts where students must reason through competing values. Argumentation, in this context, anchors this reflection, ensuring that decisions are not merely intuitive but justified. In frontier technology domains, this phase is analogous to: (1) an AI system generating concept variants during an iterative design loop, (2) a CRISPR researcher modelling off-target gene effects before experimentation, and (3) a robotics engineer simulating pathfinding algorithms prior to fabrication. Each context requires deliberate reflection, hypothesis formation, and epistemic caution.

Our STEM student designers, though working with cardboard, plastic wheels, and simple levers, are participating in the very same epistemic tradition. When individuals envision autonomous braking systems, AI-supported path optimisation, smart materials that alter stiffness under pressure, or IoT-enabled sensors that could communicate obstacle detection, they are indirectly drawing on conceptual resources from contemporary engineering. Even if these technologies are not built into the model, they shape the reasoning that informs design choices. From imagination, the team transitions to the design and construction phase, where argumentation becomes inter-locutory. Now reasoning is shared, contested, revised, and negotiated among teammates. Claims about the ramp, vehicle structure, cargo compartment, obstacle interaction, and launch mechanism must be articulated and justified using structured forms such as Claim–Evidence–Reasoning, counterclaims, rebuttals, and surrebuttals. This is where argumentation becomes a social epistemology, a way of thinking together. As teams move from imagination to design, from design to testing, and from testing to revision, each phase becomes an arena for distinct argumentative practices. This progression mirrors Fig. 5.4.

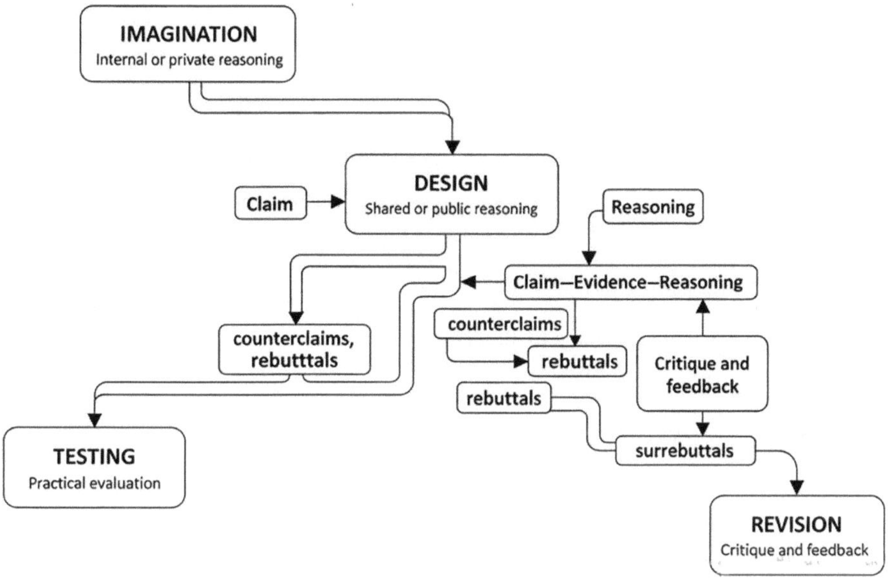

Fig. 5.4 Argumentation in engineering design

5.5.2 Designing the Sloped Ramp and Making Claims About Forces and Motion

When choosing the ramp's height and angle, team must justify: (1) If the ramp is steeper, then the potential energy increases, (2) but excessive acceleration may destabilize the vehicle, and (3) therefore an intermediate angle offers an optimal balance. Here, the team must navigate competing values, efficiency, safety, and accuracy, much like real engineers designing autonomous delivery robots or drone trajectories. Argumentation allows them to articulate tensions, defend trade-offs, and challenge unsupported assumptions. They may advance a claim such as:

Claim: *A 30° slope provides the optimal balance between gravitational acceleration and vehicle stability.*

Evidence: Calculations of gravitational force components; trial observations; friction coefficients.

Reasoning: If the slope is too steep, the vehicle may accelerate uncontrollably; if too shallow, insufficient momentum will be generated to approach the obstacle or the launch point.

Frontier technologies deepen these considerations. AI-assisted simulations, for instance, could predict the vehicle's trajectory under different slope conditions. IoT sensors could collect data in real time, producing empirical traces to support or challenge the team's assumptions. These concerns mirror the epistemic responsibility required when designing a mechanical system intended to behave predictably (Fig. 5.5).

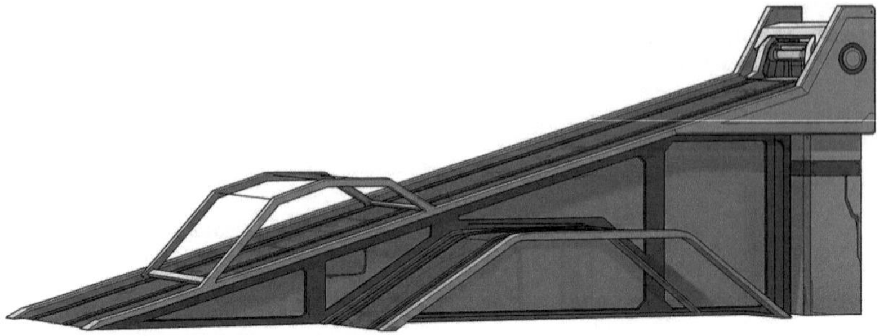

Fig. 5.5 Designing the sloped ramp for the cargo and machine

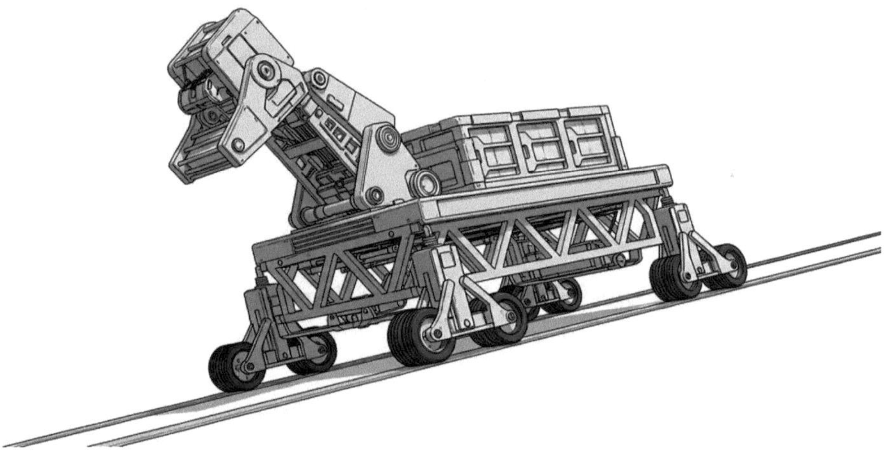

Fig. 5.6 The vehicle structure with wheels or supports

5.5.3 Designing the Vehicle Structure

Designing the body, wheels, or supports of the vehicle requires reasoning grounded in physics and engineering, such as centre of mass, friction, stability, and torque. The team must weigh competing claims:

1. *If we use larger wheels, then the vehicle will roll more smoothly over irregularities, but the increased radius may reduce control at the launch point.*
2. *If we adopt a wider wheelbase, then stability improves, but turning manoeuvrability decreases.*

Such reasoning brings in trans-locutory argumentation, where ideas migrate across imagination, design, testing, and performance. Robotics offers an illuminating analogy. In this regard, a robot's chassis design always embodies philosophical choices about stability versus agility, precision versus adaptability. The team begins to grasp that technical decisions are always epistemic decisions. For example, the structure invites as illustrated in Fig. 5.6 arguments about: (1) weight distribution, (2) wheel size, (3) support alignment, and (4) material selection. Through inter-locutory argumentation, team members challenge one another. For example, when we consider the vehicle's structure, stability, and materials, your choice of wheels will likely increase friction. It is difficult to imagine the vehicle gaining enough momentum to reach the obstacle. These exchanges develop epistemic virtues, such as patience, humility, and the willingness to revise one's stance.

5.5.4 The Cargo Compartment, Security and Flexibility

Cargo containment raises philosophical debates about security, integrity, and ethical responsibility. Argumentatively, the team must defend whether a closed, semi-closed, or open compartment best satisfies the design criteria. Figure 5.7 depicts the cargo compartment. A counterclaim may emerge: *But if the compartment is fully enclosed, then visibility of the cargo is lost, which may violate the requirement of a visible mechanism.* Other rebuttals may advance: *(1) But if we tighten the compartment too much, the launcher mechanism may jam. (2) If we loosen it, the cargo may fall prematurely.* A surrebuttal might challenge this, a team member may state that: *Transparency can be achieved via smart materials or acrylic shields.* Argumentation helps students articulate why both sides of a decision matter.

5.5.5 The Obstacle and Anticipating Uncertainty

When designing the obstacle configuration (Fig. 5.8), teams must imagine multiple future scenarios such as blocked paths, uneven surfaces, unpredictable angles. Obstacles force designers to confront uncertainty, variability, and failure points. This is fundamentally an exercise in: (1) predictive reasoning, (2) scenario modelling, (3) and contingency planning. Here, argumentation becomes probabilistic, requiring the team to justify which obstacles represent realistic challenges and which would distort the task. This reflects real-world reasoning, the same cognitive capacity required in: (1) predicting AI system failure modes, (2) modelling CRISPR ecological effects, and (3) designing autonomous robots for unpredictable environments. Argumentation allows students to justify why a particular obstacle model is realistic or worthwhile. Argumentatively, obstacles generate complex dialectical moves: (1) *If the obstacle is low, the vehicle may roll over it, but if frictional losses occur, the vehicle might fail to reach the launch point. (2) If we divert around the obstacle, then the path length increases, but stability improves.* Again, this is a moment where IoT sensors, robotic perception, and machine vision (AI) become conceptual scaffolds. By reflecting on how real-world autonomous systems detect and navigate obstacles, the team can deepen the philosophical dimension of their

Fig. 5.7 Designing the cargo compartment

Fig. 5.8 Visual representation of different obstacles

reasoning, as such, the world is not simply a set of predictable conditions but a dynamic space requiring adaptive judgment.

5.5.6 The Launch Mechanism, Precision, Force, and Counterarguments

The launcher, whether a catapult arm, spring system, or lever, requires students to reason about projectile motion, torsion, force transfer, and stability. This is the most argument-rich component. A typical argumentative exchange might unfold as:

(i) Claim: A lever mechanism gives us more predictable control.
(ii) Counterclaim: A spring system produces greater force with less structure.
(iii) Rebuttal: But springs create inconsistency unless calibrated.
(iv) Surrebuttal: Calibration is possible, and improves accuracy.

Alternatively, the team may justify their choice by stating the following:

(i) Claim: A torsion-spring mechanism provides the most reliable launch energy.
(ii) Counterclaim: But a lever arm offers better directional control.
(iii) Rebuttal: Yet lever arms introduce variability through human adjustment.
(iv) Surrebuttal: Therefore, a hybrid mechanism may balance power and precision.

Arguments here mirror debates in robotics and AI-actuated systems, that is, autonomy versus control, precision versus flexibility. Even though the team build a simple mechanical launcher, they engage implicitly with foundational engineering tensions shaping frontier technologies. This dialectical movement mirrors real engineering design review processes (Dym et al., 2005). The philosophical value lies not in who "wins" but in how each argument sharpens conceptual clarity.

Fig. 5.9 Simulation vehicle with cargo ready to launch

5.5.7 Presenting the Prototype Through Trans-Locutory Argumentation

The final phase illustrated in Fig. 5.9 requires the team to justify their designs before an audience. This mirrors engineering design reviews, ethics committee approvals, or public consultations for new technologies. Trans-locutory argumentation calls upon: (1) clarity of explanation, (2) evidence-based justification, (3) anticipation of critique, and (4) and ethical self-awareness. Presentation is not an afterthought; it is an epistemic culmination. This final phase requires rhetorical clarity, ethical justification, and evidence-based persuasion, skills essential for communicating next-generation technologies such as AI-enabled medical tools, robotics policy frameworks, or CRISPR bioethics guidelines to society. In defending their designs, students confront deeper questions: (1) Why did we choose this mechanism? (2) What assumptions guided our modelling? (3) How do we justify risks or uncertainties? These questions remind us that engineering is never merely technical; it is always epistemic and ethical.

5.5.8 Testing, Failure, Revision, and Trans-Locutory Argumentation

Testing is the moment when ideas confront reality. Up to this point, design deliberation has unfolded in the relative safety of imagination, sketches, models, and intra- or inter-locutory exchanges. However, when a team begins to test its gravity-powered cargo vehicle, rolling it down the slope, watching it encounter obstacles, observing

Fig. 5.10 Testing of the vehicle approaching an obstacle (failure)

the success or collapse of the launch mechanism, they enter a qualitatively different epistemic arena. Evidence is no longer speculative but embodied; claims are no longer hypothetical but materially challenged. In this sense, testing becomes the crucible where argumentation is forced to reconcile theory with practice. Ideas shift from design to reality, from expectation to observation, where both successes and failures provide evidence for evaluating design claims. Figure 5.10 shows the vehicle approaching an obstacle during testing and failing to overcome it. Accordingly, testing does not always confirm expectations; instead, it often reveals failure, which becomes the starting point for deeper inquiry and revision.
We have learned that failure, in this phase, is neither accidental nor undesirable. Failure is epistemically productive. It exposes hidden assumptions, faulty inferences, weak evidence, or overly confident interpretations. When the vehicle tips to one side, or when the cargo fails to launch toward the target zone, the team must confront the tension between what they believed and what actually transpired. This confrontation is not merely technical; it is argumentative. The team must articulate what they think went wrong, provide reasons for their interpretations, respond to competing explanations from teammates, and negotiate which line of evidence should guide revision. In this respect, failure becomes a dialogue partner that speaks through friction, instability, or misalignment, demanding justification. The team must revisit their reasoning:

1. Our claim about stability was insufficient because the vehicle veered left.
2. Our evidence for the slope angle must be revised.
3. Our rebuttal regarding the wheel radius no longer holds.

The testing phase becomes the philosophical heart of argumentation, as it forces humility, intellectual honesty, and responsiveness to evidence, precisely the dispositional qualities needed in next-generation technological world where models, predictions, and algorithms must continually be validated against reality. This process mirrors the dynamics of trans-locutory argumentation, the public defence of design reasoning before an external audience. In the testing phase, even if the audience consists only of peers observing from across the room, the team becomes accountable to perspectives beyond their own. They must explain why the ramp angle proved insufficient, why the wheel alignment introduced drag, or why the launch mechanism failed under stress. Their argumentation shifts from internal justification ("I think this because…") and group negotiation ("We believe this because…") to public accountability ("Here is why this happened, based on the evidence available to us…"). Trans-locutory argumentation is thus inseparable from the epistemic humility required to revise one's design.

This humility becomes even more crucial when we consider the growing influence of frontier technologies. Contemporary engineering is increasingly shaped by fields such as artificial intelligence, robotics, CRISPR, and the Internet of Things, domains defined by unpredictability, emergent behaviour, and complex system interactions. In these domains, failure during testing is not a mere inconvenience but an ethical signal. A self-driving car misinterpreting an obstacle, a genetic-editing algorithm producing unintended modifications, or an IoT network reacting

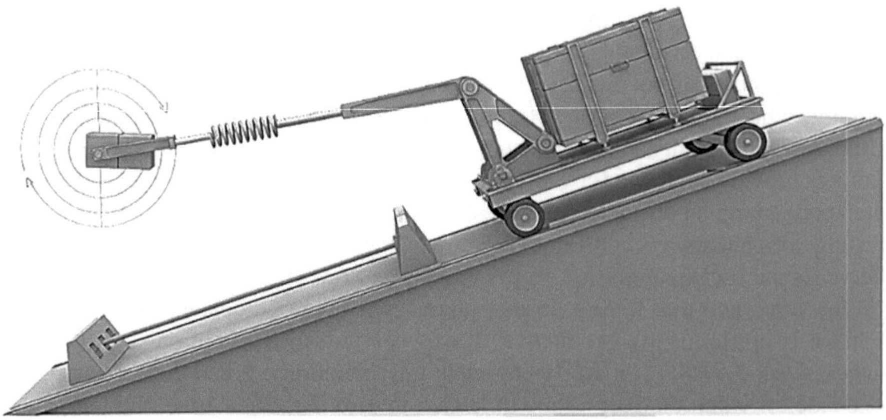

Fig. 5.11 Re-testing of the simulated vehicle overcoming the obstacle (successful)

unpredictably to environmental noise: these failures necessitate robust argumentative practices that integrate empirical evidence, ethical foresight, and interdisciplinary reasoning. By engaging with failure in the controlled context of a learning environment design task, students rehearse the same argumentative virtues needed to navigate these high-stakes technological landscapes.

Revision follows naturally from this encounter with failure. However, revision is not simply the mechanical act of altering a design. It is the culmination of argumentation. The team must weigh competing explanations, reconcile conflicting interpretations of evidence, and decide what matters most for improving the design. Should they prioritize stability over speed? Precision over distance? Reliability over innovation? These decisions are philosophical as much as they are technical, because they reflect the underlying values the team brings to the design process. Through revision, argumentation acquires a temporal dimension. Ideas are tested, fail, are defended, critiqued, and reshaped. Claims evolve, evidence is reinterpreted, warrants shift in strength, and new counterarguments emerge. Students begin to see that design is not linear but recursive, and that the recursive movement is fuelled by the dynamic interplay of reasoning, critique, creativity, and ethical judgment.

The moment captured in Fig. 5.11 marks a decisive epistemic turning point in the design process. After earlier failed attempts in which the prototype stalled, tipped, or lost momentum when approaching the obstacle, the team engaged in a cycle of reflection, argumentation, and revision, revisiting assumptions about friction, ramp angle, wheel alignment, load distribution, and force transfer through the launch mechanism. What we see in this successful re-test is not merely improved engineering performance but the material expression of improved reasoning. As the vehicle approaches the obstruction, its motion reveals that previously competing claims have now been reconciled through evidence-based justification. The wheels maintain stability, the suspension absorbs the rise, and the vehicle transitions over the obstacle with controlled momentum. This moment dramatizes how argumentation reshapes design: the successes and failures of earlier experiments became data

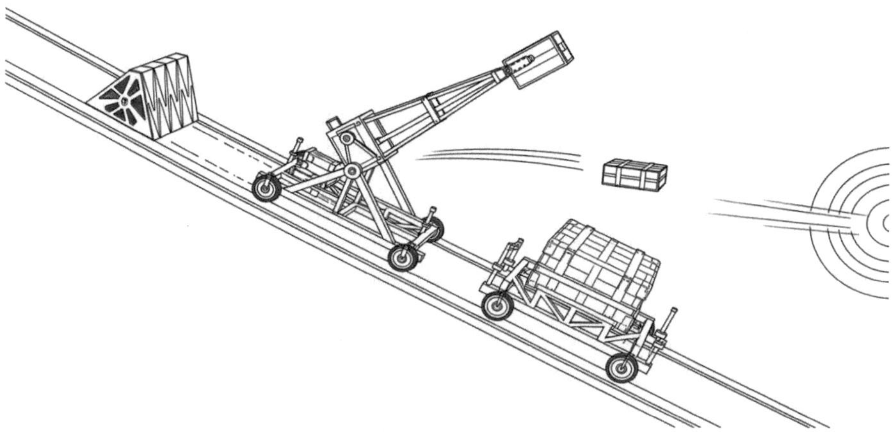

Fig. 5.12 Successful launch of cargo-loaded vehicle

points around which students constructed claims, counterclaims, warrants, and rebuttals. Importantly, the re-test invites reflection on the broader engineering cultures shaped by frontier technologies. In fields such as autonomous robotics, AI-guided navigation, and IoT-enabled mobility systems, re-testing is fundamental, not because machines must be perfected, but because reasoning about complex systems requires iterative confrontation with uncertainty. By observing the vehicle overcoming the obstacle in this final test, the team gained experiential insight into the recursive, argumentative, and ethically accountable processes that underpin responsible technological innovation.

Finally, trans-locutory argumentation reaches its fullest expression when the team presents its revised design to an external audience, classmates, teachers, assessors, or community members. In this culminating moment, the team must articulate not only what they built but why they built it this way. They must explain how failures shaped their reasoning, how evidence guided each decision, and how alternative possibilities were considered and ruled out. In doing so, they enact the core intellectual competencies required of future engineers working in societies transformed by frontier technologies: the ability to reason publicly, justify decisions transparently, and demonstrate moral accountability for the systems they help bring into the world. In this phase, argumentation becomes a lived practice, a synthesis of cognition, collaboration, embodied experience, and ethical imagination. It teaches students that knowledge is not something one possesses but something one must constantly defend, refine, and renegotiate. Testing and revision thus illuminate a profound educational truth. In STEM, as in life, failure is not the opposite of success; it is the engine of understanding. After revision, the design is re-tested and was successfully launched to the desired target.

Figure 5.12 depicts the culmination of the team's design journey. A fully loaded vehicle descending the track, encountering the blockade, activating the launch mechanism, and propelling its cargo accurately toward the target zone. This is more than a mechanical triumph, it is the embodiment of coordinated cognitive, argumentative, and ethical work. The mechanics of the launch unfold in a sequence shaped by prior disagreements, revisions, and negotiated reasoning. Earlier debates about spring tension, lever length, pivot placement, weight distribution, and the angle of release reappear here as refined, integrated design elements. The successful launch demonstrates that the team's claims about the optimal configuration of the mechanism have now been validated through observable evidence. What had once existed only as intra-locutory hypotheses or inter-locutory deliberations becomes, in this moment, trans-locutory knowledge, publicly defensible, empirically grounded, and communicable to an external audience.

This figure also provides an opportunity to engage philosophically with the ethics of frontier technologies. Launch mechanisms in contemporary engineering echo real-world systems such as robotic manipulators, drone payload delivery systems, AI-assisted targeting mechanisms, and even CRISPR gene-delivery vectors in biotechnology. In each case, accuracy, risk, and responsibility intertwine. The successful cargo launch invites us to consider how design decisions, however small in a classroom project, implicitly rehearse the principles of precision, safety, and accountability demanded by cutting-edge technological systems. Thus, the scene captured in Fig. 5.12 is not merely the final step of a project. It represents the culmination of a pedagogical philosophy, showing that argumentation, when infused into design practice, cultivates the intellectual virtues necessary for navigating a human-centered technological future.

This exercise reveals that engineering design is never purely technical. It compels us to confront ethical questions: (1) How do we reason responsibly in the face of uncertainty? (2) How do we justify design decisions that may have real-world consequences? (3) What counts as sufficient evidence when claims and counterclaims compete? (4) How do frontier technologies influence not just what we build, but how we think? As Iwuanyanwu (2023) argues, argumentation cultivates epistemic virtues, critical friendship, reflective judgment, and ethical imagination. Through the engineering design task, we have illustrated how even a seemingly simple challenge, constructing a gravity-powered cargo vehicle, becomes a microcosm of the intellectual and moral demands placed upon learners in the age of frontier technologies. At each stage of the design cycle, argumentation provided the conceptual scaffolding through which ideas were imagined, challenged, refined, and defended. What emerged was not merely a prototype vehicle, but a demonstration of how structured reasoning acts as the connective tissue linking scientific knowledge, engineering judgement, ethical responsibility, and collaborative intelligence. In this sense, argumentation is best seen as a cognitive ecology, a space in which claims, counterclaims, evidence, values, and emotions interact dynamically to shape understanding.

Crucially, the argumentation processes embedded in the design task resonate with the broader epistemic challenges confronting contemporary STEM fields. As we encounter powerful technologies such as artificial intelligence, CRISPR gene editing, autonomous robotics, and expansive IoT ecosystems, we find ourselves navigating controversies that cannot be resolved by technical expertise alone. These controversies demand human judgment that is informed by evidence but guided by ethical and philosophical reflection. For instance, when the teams debate the optimal angle of the ramp or the most reliable launch mechanism, they are also learning how engineers evaluate risk in self-driving cars or how geneticists weigh potential benefits against unintended consequences. When teams contest interpretations of stability, safety, or efficiency, they rehearse the kinds of argumentation required in real-world contexts where stakeholders disagree, values conflict, and uncertainties persist. These analogies are not metaphors. They represent the continuity between classroom argumentation and the lived realities of technologically mediated societies.

Moreover, as the team transition from intra-locutory (self) reasoning to inter-locutory (collaborative) negotiation and finally to trans-locutory (public) defense, they enact a progression that parallels the lifecycle of technological innovation itself. Ideas begin as internal hypotheses, become collaboratively shaped through critique, and enter the public sphere where they must withstand scrutiny, debate, and accountability. We witness here a profound educational insight that shows argumentation cultivates both intellectual autonomy and moral agency. It equips teams not only to solve problems but to justify solutions responsibly and transparently. Thus, argumentation emerges as an indispensable tool for developing the kind of STEM thinker needed in our century: reflective, creative, ethical, and attuned to the social consequences of innovation. As we move into the next chapter, we will deepen this analysis by examining how argumentation supports advanced cognitive development, metacognitive awareness, and the cultivation of ethical imagination in technologically intensive environments. In doing so, we reinforce a central claim of this book that the future of STEM depends not on our capacity to build smarter machines, but on our ability to nurture wiser humans.

5.6 Chapter Summary

In this chapter, we examined argumentation as a foundational epistemic and pedagogical practice within STEM education, particularly in a world increasingly shaped by frontier technologies such as artificial intelligence, CRISPR gene editing, robotics, and expansive IoT ecosystems. We established that argumentation operates far beyond the exchange of claims; it is a dynamic, culturally embedded, and ethically charged form of reasoning that prepares learners to interrogate evidence, evaluate competing interpretations, and deliberate responsibly in technologically complex environments. We began by distinguishing between arguments and argumentation before tracing the evolution of argumentative theory from classical philosophical traditions to Toulmin's Argument Pattern (TAP) and the Contiguity Argumentation Theory (CAT). While TAP illuminates the structural elements of arguments, CAT highlights the shifting interplay among logical, non-logical, and value-laden forms of reasoning, revealing how individuals navigate competing worldviews in socio-scientific contexts. These frameworks, taken together, underscore the need to understand argumentation as both a cognitive and cultural activity.

We then illustrated the practical utility of argumentation through an engineering design demonstration involving a gravity-powered cargo vehicle. Across the phases of imagination, design, testing, failure, revision, and public presentation, argumentation emerged as the epistemic engine that made reasoning visible, improvable, and accountable. Students engaged in intra-locutory, inter-locutory, and trans-locutory reasoning, learning to justify design decisions, challenge assumptions, interpret failure productively, and defend their choices before an external audience. Ultimately, the chapter argues that STEM education must foreground argumentation to cultivate the intellectual virtues necessary for responsibly engaging with frontier technologies. By learning to reason critically, collaborate dialogically, and deliberate ethically, students develop not only scientific literacy but the broader capacities required to contribute meaningfully to a human-centered technological future.

5.6.1 Looking Ahead to Chap. 6

In Chap. 5, we established argumentation as a foundational epistemic and pedagogical practice in STEM education, showing how individuals develop the capacity to justify claims, evaluate evidence, and deliberate responsibly in technologically complex contexts. We demonstrated that argumentation makes reasoning visible, open to critique, and subject to refinement, enabling students and professionals to engage not only with what they know but also with how they know and why their knowledge matters. Yet, as we have seen, argumentation depends on more than the ability to structure claims and counterclaims; it relies on the quality of the evidence available and the scientific practices through which such evidence is generated, interpreted, and communicated.

In Chap. 6, we extend this inquiry by examining the scientific practices that make meaningful argumentation possible. We turn our attention to the foundational practices of observation, classification, measurement, prediction, inference, and communication, showing how these practices equip learners with the tools needed to engage rigorously with scientific phenomena. We then build on these foundations to explore more advanced competencies, including designing investigations, controlling variables, conducting experiments, interpreting data, and constructing explanations. Through this progression, we illustrate how scientific practices form an interconnected system that supports the construction, evaluation, and refinement of knowledge.

As we move forward, we also deepen our focus on argumentation by situating it within the broader ecosystem of scientific practice. We show how argument from evidence emerges naturally when learners engage systematically in scientific inquiry, and how these practices strengthen their capacity for informed decision-making and ethical reasoning. In doing so, we invite readers to see scientific practices not as isolated skills, but as integral to cultivating scientifically literate individuals who can reason responsibly, communicate transparently, and participate thoughtfully in shaping a human-centered technological future.

5.7 Reflection Questions

1. Conceptual Understanding
   (i) How does argumentation differ from arguments, and why is this distinction important for STEM education?
   (ii) In what ways do TAP and CAT complement and challenge one another in explaining real-world reasoning?
2. Application of Theory
   (i) Using an example from AI, CRISPR, robotics, or IoT, identify a situation where logical, non-logical, and value-laden arguments might all emerge.
   (ii) How do worldview elements (dominant, suppressed, assimilated, emergent, equipollent) manifest in socio-scientific debates?
3. Design and Engineering Reasoning
   (i) Reflect on the design cycle presented in the engineering task. At which stage do you think argumentation is most critical, and why?
   (ii) How does failure function as an epistemic resource in engineering or scientific inquiry?
4. Ethical and Epistemic Reflection
   (i) What does it mean to exercise "epistemic humility" in the context of frontier technologies?
   (ii) How might argumentation help prevent uncritical technological optimism in STEM fields?
5. Meta-Cognitive Reflection
   (i) Consider your own reasoning practices. How do you engage in intra-locutory, inter-locutory, and trans-locutory argumentation when solving complex problems?

Bibliography

Asterhan, C., & Schwarz, B. (2007). The effect of monological and dialogical argumentation on concept learning in evolution theory. *Journal of Education Psychology, 99*(3), 626–639.
Britannica. (2008). *Argument*. Encyclopaedia Britannica.

Dym, C. L., Agogino, A. M., Eris, O., Frey, D. D., & Leifer, L. J. (2005). Engineering design thinking, teaching, and learning. *Journal of Engineering Education, 94*(1), 103–120. https://doi.org/10.1002/j.2168-9830.2005.tb00832.x

Erduran, S., & Park, W. (2023). Argumentation in physics education research: Recent trends and key themes. In M. F. Taşar & P. R. L. Heron (Eds.), *The international handbook of physics education research: Learning physics* (pp. 16–32). AIP Publishing.

Erduran, S., Simon, S., & Osborne, J. (2004). TAPping into argumentation: Developments in the application of Toulmin's Argument Pattern for studying science discourse. *Science Education, 88*(6), 915–933. https://doi.org/10.1002/sce.20012

Iwuanyanwu, P. N. (2020). Nature of problem-solving skills for 21st century STEM learners: What teachers need to know. *Journal of STEM Teacher Education, 55*(1), 27–40. https://doi.org/10.30707/JSTE55.1/MMDZ8325

Iwuanyanwu, P. N. (2023). When science is taught this way, students become critical friends: Setting the stage for student teachers. *Research in Science Education, 53*(6), 1063–1079. https://doi.org/10.1007/s11165-023-10122-9

Iwuanyanwu, P. N., & Ogunniyi, M. B. (2020). Effects of dialogical argumentation instructional model on pre-service teachers' ability to solve conceptual mathematical problems in physics. *African Journal of Research in Mathematics, Science and Technology Education, 24*(1), 121–141. https://doi.org/10.1080/18117295.2020.1748325

Kuhn, D., & Udell, W. (2007). The development of argument skills. *Child Development, 78*(4), 1241–1257. https://doi.org/10.1111/j.1467-8624.2007.01062.x

Lubben, F., Sadeck, M., Scholtz, Z., & Braund, M. (2010). Gauging students' untutored ability in argumentation about science-related issues. *Research in Science Education, 40*(2), 227–248. https://doi.org/10.1007/s11165-008-9106-0

Ogunniyi, M. B. (2007a). Teachers' stances and practical arguments regarding a science indigenous-knowledge curriculum: Part 1. *International Journal of Science Education, 29*(8), 963–986. https://doi.org/10.1080/09500690600931020

Ogunniyi, M. B. (2007b). Teachers' stances and practical arguments regarding a science indigenous-knowledge curriculum: Part 2. *International Journal of Science Education, 29*(10), 1189–1207. https://doi.org/10.1080/09500690600931046

Osborne, J. F. (2019). Not "hands on" but "minds on": A response to Furtak and Penuel. *Science Education, 103*, 1280–1283. https://doi.org/10.1002/sce.21543

Schwarz, B. B., Rapanta, C., & Asterhan, C. S. (2017). Argumentation in education. In A. A. Ioannou, S. Papadopoulou, & T. Mikropoulos (Eds.), *Handbook of educational technology research* (pp. 287–302). Springer. https://doi.org/10.1007/978-3-319-60013-0_20

Simon, S., & Johnson, S. (2008). Professional learning portfolios for argumentation in school science. *International Journal of Science Education, 30*(12), 1701–1721. https://doi.org/10.1080/09500690701579572

Toulmin, S. (2003). *The uses of argument*. Cambridge University Press.

Van Eemeren, F. H., Grootendorst, R., & Kruiger, T. (1987). *Handbook of argumentation theory*. Foris Publications.

Chapter 6
Scientific Practices for Developing STEM Thinking, Decision-Making, and Argumentation

6.1 Introduction

We imagine a young engineer adjusting the sensors on a prototype, testing a design under uncertain conditions. Nearby, a scientist debates with colleagues over conflicting data, weighing evidence, questioning assumptions, and defending interpretations. Across laboratories, classrooms, and field sites, STEM professionals are not merely following procedures, they are reasoning, observing, experimenting, and arguing in real time. These moments reveal that scientific practice is more than a set of techniques; it is a way of thinking, a disciplined approach to navigating uncertainty and making decisions responsibly. At the heart of this practice are foundational skills; observation, classification, measurement, prediction, inference, and communication, that provide the tools for engaging meaningfully with phenomena. These skills then support more advanced competencies, from designing investigations and controlling variables to interpreting data and constructing explanations. Together, they form an interconnected system that enables learners and professionals to generate, evaluate, and refine knowledge.

Within this system, argumentation becomes indispensable. Reasoning from evidence is not optional; it is the very mechanism that makes scientific practice accountable, reflective, and ethically grounded. By engaging deeply with these practices, STEM thinkers develop not only technical competence but the capacity to reason critically, make informed decisions, and act responsibly. Scientific practice, in this sense, is not simply about doing science; it is about learning how to think responsibly within conditions where certainty is never fully guaranteed. To understand this more clearly, we must begin by recognising that scientific practice is fundamentally epistemic (Osborne, 2019). It exists to transform uncertainty into justified understanding. Every investigation, every model, and every explanation represents an attempt to move from what is not yet known to what can be reasonably defended. This movement requires more than observation. It requires judgment. It

P. N. Iwuanyanwu, *Empowering STEM Thinkers Through Argumentation*, Contemporary Trends and Issues in Science Education 69,
https://doi.org/10.1007/978-3-032-27906-4_6

requires decisions about what counts as relevant evidence, which methods are appropriate, and whether conclusions are warranted by the available data. As we begin to reflect on our own engagement with STEM inquiry, we may find ourselves asking: *How do we know that a conclusion is reliable? What distinguishes evidence from mere observation? At what point does explanation become justification?* These questions signal our entry into scientific practice as a form of disciplined reasoning rather than mechanical procedure.

As we move further into this chapter, we begin to see that scientific practices form an integrated architecture of inquiry. What initially appears as a sequence of steps reveals itself as a dynamic interplay of questioning, investigation, modelling, evaluation, and argumentation. We find ourselves formulating questions that expose uncertainty, designing investigations that render phenomena observable, and interpreting evidence that rarely speaks for itself. At each point, we are required to make judgments. We must decide whether evidence is sufficient, whether patterns are meaningful, and whether alternative explanations have been adequately considered. In doing so, we begin to recognise that scientific practice is not merely about producing answers, but about ensuring that answers are justified.

This realisation becomes even more significant as we consider the contemporary technological landscape. Increasingly, STEM professionals work alongside computational systems capable of analysing vast quantities of data and generating predictive models. These systems extend our cognitive reach, but they do not replace our epistemic responsibility. Instead, they deepen it. We must evaluate the assumptions embedded in algorithms, interpret the outputs they produce, and determine whether their conclusions are valid. We are therefore compelled to confront another fundamental question: *Where does responsibility for knowledge reside when machines participate in its production?* The answer returns us to scientific practice itself. It is through disciplined investigation, critical evaluation, and reasoned argumentation that knowledge retains its integrity.

As we proceed, we will see that scientific practices do more than produce knowledge. They cultivate habits of mind. They teach us to tolerate uncertainty without surrendering to it, to question without dismissing, and to conclude without assuming finality. In this way, we see that scientific practice is inseparable from the intellectual formation that Popper (2002) described in Conjectures and Refutations. Engaging with evidence, testing ideas, and refining explanations is not just a method, it is how we cultivate the critical, reflective, and responsible thinking that shapes STEM professionals. It thus stands to reason that scientific practice shapes not only what we know, but how we come to know. By the end of this chapter, we will recognise scientific practice not as a rigid method, but as an epistemic discipline, one that enables us to think clearly, decide responsibly, and argue persuasively in a world increasingly shaped by complex technologies and uncertain futures.

6.2 Designing and Implementing STEM Investigations

As we enter the domain of scientific investigation, we begin to encounter inquiry in its most deliberate form. Investigation represents the moment where curiosity becomes structured, where uncertainty becomes examinable, and where questions begin their transformation into knowledge. It is here that scientific practice becomes visible. We move from wondering about phenomena to designing ways of observing them systematically. We shift from speculation to evidence. At first glance, investigation may appear to be a technical exercise, a matter of selecting instruments, collecting data, and applying analytical procedures. Yet as we look more closely, we begin to recognise that investigation is fundamentally an act of reasoning. Before any measurement is taken, decisions must be made. We must determine what to investigate, which variables matter, and how observations should be interpreted. These decisions are not given to us. They require judgment. They require us to reflect on what we are trying to understand and why it matters.

As we engage in this process, we may begin to notice something important. Investigation does not eliminate uncertainty immediately. Instead, it allows us to engage uncertainty productively. Each observation provides partial insight. Each measurement offers a piece of a larger pattern. Gradually, through careful design and systematic observation, uncertainty becomes structured. We begin to see relationships where previously there was only ambiguity. Investigation, in this sense, is not the removal of uncertainty, but its disciplined transformation into evidence. This process also reveals that investigation is inherently interpretive.

Data do not explain themselves. They must be interpreted within conceptual frameworks that give them meaning. We must ask ourselves: *What does this pattern suggest? Could there be alternative explanations? Have we overlooked variables that influence the outcome?* These questions remind us that investigation is not separate from reasoning; it is reasoning made visible through structured engagement with the world. As STEM professionals, we therefore come to understand that designing investigations is not merely about following established procedures. It is about exercising epistemic responsibility. It requires us to ensure that our methods are coherent, that our evidence is sufficient, and that our conclusions are justified. In this way, investigation becomes more than a technical task. It becomes a disciplined practice through which knowledge is constructed, evaluated, and defended.

6.2.1 *Formulating Investigable Questions*

Our entry into investigation begins with a question. Yet not every question can sustain scientific inquiry. Some questions invite speculation but resist empirical examination. Others conceal assumptions that predetermine their answers. To engage in disciplined investigation, we must learn to formulate questions that open inquiry rather than close it. As we begin to reflect on this process, we may notice that formulating a scientific question requires us to confront our own uncertainty. We must identify what we do not yet understand and articulate it in a form that can be examined systematically. This act alone represents a significant intellectual shift. It requires us to move from passive curiosity to active inquiry. We are no longer merely wondering; we are preparing to investigate.

In doing so, we must also learn to discipline our thinking. An investigable question must be clear enough to guide observation, yet open enough to allow evidence to shape the answer. It must identify relationships that can be examined without assuming conclusions in advance. As we refine our questions, we may find ourselves asking: *Can this phenomenon be measured? What evidence would help answer this question? Are there variables that must be controlled?* These reflections help transform vague curiosity into structured inquiry. This process also deepens our awareness of the relationship between questioning and argumentation. A well-formulated question anticipates the need for justification. It creates the conditions under which claims can be supported by evidence and evaluated critically. In this way, the question becomes the first step in the construction of a reasoned argument. It establishes the epistemic pathway along which investigation will proceed.

As we develop this capacity, we begin to recognise that questioning is not a preliminary stage to be discarded once investigation begins. It remains active throughout inquiry. New evidence may challenge our initial assumptions. Unexpected patterns may require us to refine our questions. Investigation and questioning therefore evolve together, each informing and strengthening the other. Through this process, we come to see that the ability to formulate investigable questions represents one of the most important scientific practices we can develop. It enables us to engage uncertainty constructively, design meaningful investigations, and construct knowledge that can be defended through evidence and reasoning.

Equally important is the dialogic nature of question formulation. Scientific questions emerge through processes of critique, refinement, and collaborative reasoning. Practitioners interrogate assumptions, challenge conceptual ambiguities, and consider alternative framings. This process strengthens the epistemic integrity of inquiry by ensuring that questions are clear, coherent, and investigable. Research in computational and mathematical thinking confirms that the formulation of meaningful problems is central to developing higher-order reasoning and scientific understanding (Lee et al., 2023; Lockwood & Mørken, 2021).

For educators and emerging professionals, the ability to formulate investigable questions represents a foundational scientific practice. It cultivates intellectual discipline, epistemic awareness, and argumentation. Through questioning, practitioners learn to identify uncertainty, clarify assumptions, and structure inquiry in ways that support reliable knowledge construction. Table 6.1 illustrates how carefully formulated investigable questions guide meaningful scientific exploration. These examples demonstrate that rigorous questioning strengthens evidence-based reasoning, improves investigative coherence, and enhances the quality of both scientific understanding and professional practice.

Table 6.1 Sample of investigable questions illustrating robust STEM inquiry practices

Principle of Robust STEM Questioning	Investigable Questions
1. Revealing uncertainty and resisting premature closure	• How do variations in local soil properties influence the stability and performance of school or community structures during seismic activity? • In what ways do students' initial mental models of energy transfer shape their reasoning when interpreting experimental results? • What factors might explain differential uptake and effectiveness of AI-supported instructional feedback across diverse STEM classrooms?
2. Avoiding embedded conclusions or unexamined assumptions	• Under what instructional conditions does inquiry-oriented teaching support deeper mechanistic reasoning in STEM subjects compared to more directive approaches? • How are distinct patterns of digital device use related to measurable changes in learner engagement and attentional focus during STEM laboratory activities?
3. Making variables, assumptions, and contextual constraints explicit	• How does the availability and quality of laboratory equipment influence students' capacity to design valid investigations within resource-constrained schools or community training centres? • To what extent do different teacher questioning strategies impact students' depth of explanation and argumentation during collaborative STEM problem-solving?
4. Ensuring ethical and scientific grounding in an AI-enhanced environment	• How do educators and practitioners evaluate the ethical appropriateness of AI-generated instructional suggestions, particularly regarding data privacy, contextual relevance, and learner autonomy? • What criteria do STEM researchers use when assessing whether AI-generated hypotheses align with disciplinary standards of rigour, feasibility, and methodological soundness?
5. Inviting multiple plausible explanations	• What mechanisms may account for differences in student motivation, persistence, or conceptual understanding when learning through virtual simulations versus hands-on experimentation? • What system-level, social, or environmental factors could explain variability in community responses to early-warning disaster technologies implemented by STEM professionals?
6. Specifying observable or inferable phenomena	• How do students' observable reasoning strategies evolve across successive cycles of modelling, prediction, and revision within a STEM inquiry activity? • What identifiable patterns emerge in educators' feedback practices as they integrate AI-assisted assessment tools into their teaching routines?
7. Demonstrating refinement through dialogic, collaborative professional engagement	• How can the construct of "STEM investigative curiosity" be more precisely operationalised for pre-service and in-service teachers engaged in inquiry-based practice? • Which aspects of students' experimental design proposals require further conceptual clarification to ensure methodological coherence?

6.2.2 Selecting Appropriate Methods

Whether an investigation is experimental, observational, computational, qualitative, or model-based, the selection of methods is always an act of judgement. Choosing a method is, in effect, a form of argumentation. That is to say, every methodological decision implicitly advances claims about what kinds of evidence are most relevant, which variables are most consequential, what limits our ability to generalise, and which tools or techniques are sufficiently valid for the task at hand. A robotics engineer, for example, must decide whether a perception system should rely on LiDAR, machine-vision cameras, or a sensor-fusion approach, each choice signalling different assumptions about accuracy, environmental robustness, and computational constraints. Similarly, a bioengineer working with CRISPR technology may judge that potential off-target effects require whole-genome sequencing rather than targeted assays, particularly when the ethical implications of the modification are significant. In this way, methodological choices function as explicit warrants within the broader structure of scientific argumentation, shaping not only how data are gathered but how claims are justified. Mostly, practitioners consider:

(i) Feasibility: What can be measured, simulated, or observed with available technologies?
(ii) Resolution: What level of precision is required to discriminate between competing explanations?
(iii) Ethical constraints: What forms of data collection are permissible or responsible?
(iv) Epistemic affordances: Which methods are capable of generating the kind of evidence needed for the claim?

In a technologically advanced landscape, method selection increasingly involves automated sensors, virtual laboratories, AI-supported modelling tools, or remote data acquisition systems. Yet professional judgement remains central in a sense that automation can augment measurement, but it cannot determine whether those measurements are meaningful.

6.2.3 Establishing Controls and Clarifying Causality

Controls, baselines, and comparison conditions remain foundational to causal reasoning in STEM practice (Akiri et al., 2020). However, in many real-world contexts, seismic events, ecological systems, human behaviour, strict controls are impossible. Investigations in STEM rarely produce perfect certainty, and professional inquiry requires continuous attention to the conditions under which evidence becomes trustworthy. This includes establishing appropriate controls when causal relationships are under question, documenting the assumptions embedded in models, instruments, or computational pipelines, and recognising the inherent limits of sampling strategies or sensor accuracy. It also involves anticipating how different sources of uncertainty may distort interpretations, particularly when data are incomplete, noisy, or filtered through automated systems. Automation does not eliminate uncertainty; in many cases, it amplifies it.

On the other hand, AI models can introduce epistemic opacity, IoT sensor networks often generate fragmented or unreliable data streams, and genetic engineering techniques may yield unintended yet mechanistically plausible phenotypic outcomes. For these reasons, scientific practice must cultivate an "argumentative habit", a disciplined commitment to making uncertainty visible, interpretable, and ethically explicit. In this light, professionals therefore:

(i) use statistical, computational, or comparative strategies to infer causal relationships;
(ii) document the limitations of their causal claims;
(iii) consider alternative explanations proactively; and
(iv) justify why their methodological decisions are appropriate for the context.

This transparency renders investigative design itself a form of argumentation where the practitioner defends why a particular design yields trustworthy inferences. Attending to uncertainty and limitations is equally paramount. Uncertainty is not a flaw in scientific work; it is a structural condition of the practice itself. Professionals learn to treat uncertainty as something to be examined rather than eliminated. They quantify it when possible, through measures such as error margins, confidence intervals, or sensitivity analyses. They remain attentive to the limits of their instruments, sampling strategies, and model assumptions, and they work to distinguish genuine patterns from noise produced by data or methodological constraints. Just as importantly, they communicate uncertainty in ways that enable responsible decision-making, ensuring that the limits of knowledge are rendered visible rather than implicit. In educational contexts, cultivating students' sensitivity to these epistemic constraints prepares them for the realities of professional STEM environments (Akiri et al., 2020), where significant decisions must often be made in the presence of ambiguity and incomplete information (Iwuanyanwu, 2020). Such sensitivity is equally important for the professional development of STEAM educators (Mafokwane, 2026).

6.2.4 Integrating Technology Responsibly

Just as frontier technologies open new possibilities for what can be imagined, so too do specific innovations, AI-enhanced diagnostic systems, simulation environments, automated data-logging tools, and machine-learning classifiers, expand what we can investigate. These technologies extend our ability to observe, measure, and interpret phenomena, while simultaneously raising new questions and opportunities for rigorous reasoning. In this way, technological advancement and scientific practice move together, shaping both the scope of inquiry and the quality of argumentation that underpins responsible STEM work. Today's investigations frequently integrate digital sensors and high-resolution instrumentation, computer simulations and digital twins, AI-assisted data curation and analysis, robotics systems capable of high-precision operations, and automated statistical modelling tools. Yet even as these technologies expand what can be measured, modelled, or inferred, professional judgement cannot be outsourced.

Practitioners must still evaluate whether automated outputs are warranted, determine when an algorithm's recommendation should be overridden, and distinguish high-confidence predictions from artefacts produced by the data pipeline itself. They must also ensure that ethical considerations, privacy, safety, fairness, and the broader social consequences of scientific decisions, guide their investigative choices. In this way, the responsible integration of technology becomes an epistemic skill in its own right, requiring discernment, interpretive sensitivity, and an explicit awareness of the limits of automation. Yet they introduce new intellectual demands:

(i) Algorithmic transparency: What does the tool measure? How does it determine patterns?
(ii) Data provenance: Are datasets reliable, representative, and ethically sourced?
(iii) Interpretive judgement: Do AI-generated outputs warrant acceptance, or do they require sceptical scrutiny?

Distributed intelligence does not replace scientific reasoning. Instead, it shifts the practitioner's role toward evaluating technological outputs, rather than merely producing them.

6.2.5 *Documenting Reasoning and Methodological Justification*

A well-designed investigation does not merely generate data; it anticipates critique. Its methods must be defensible, its evidence clearly relevant to the claims it aims to support, its assumptions explicit and open to revision, and its limitations acknowledged rather than concealed. In this respect, investigative design becomes the first stage of argumentation. Every decision, from the way a question is framed to the selection of methods, the integration of technology, and the interpretive stance adopted, functions as a claim that must ultimately be warranted through justification. High-quality investigations articulate:

(i) why particular methods were selected,
(ii) how decisions were made,
(iii) what alternatives were considered,
(iv) how limitations were mitigated, and
(v) how evidence supports emerging claims.

These criteria transform investigation into a reasoned argument, a narrative of inquiry that makes the researcher's epistemic moves visible. In professional practice, such transparency is essential for peer review, regulatory evaluation, and interdisciplinary collaboration. This is because scientific investigations do not occur in isolation. They are inherently social and value-laden, drawing on collaboration, critique, and shared reasoning within professional communities. In argumentation-rich STEM environments, classrooms, laboratories, or industry settings, ideas are refined through productive disagreement. Practitioners and emerging professionals challenge one another's assumptions, interrogate the robustness of evidence, and justify methodological choices. These practices mirror the demands of real-world STEM work (e.g., Dym et al., 2005; Ogunniyi & Iwuanyanwu, 2024; Peretz et al., 2024; Simon & Johnson, 2008). This collective engagement strengthens the quality of inquiry, ensuring that conclusions are not merely asserted but defended through transparent and accountable reasoning.

Equally important, every investigation carries ethical implications. The questions we choose to pursue, the methods we select, the data we privilege, and the interpretations we advance all shape how scientific knowledge enters and influences the world. Investigations can affect communities, ecosystems, and societal futures, meaning that scientific design is never solely a technical act, it is a moral and civic one. To design an investigation, therefore, is also to design a set of responsibilities: responsibilities to evidence, to professional integrity, and to the broader human and environmental contexts within which STEM knowledge will ultimately be applied.

6.3 Thinking, Reasoning, Interpreting Data, and Arguing in STEM

In professional STEM work, data are not passive objects waiting to be read, they are artefacts that demand interpretation, judgement, and epistemic sensitivity. Data may be collected by sensors, simulations, instruments, or AI-driven analytics, but they do not speak for themselves. Their meaning emerges only when a practitioner engages with them critically, discerning patterns, identifying anomalies, and examining what those patterns reveal about the underlying phenomena. Effective STEM reasoning therefore begins not with description, but with interpretation: the ability to look beyond surface outputs and recognise the deeper relationships, dependencies, and causal structures that data may be signalling. Interpreting data requires selecting forms of representation that clarify rather than obscure insight. Graphs, models, statistical summaries, or computational visualisations act as cognitive tools that bring structure to complexity. Yet each representation foregrounds certain aspects of the data while downplaying others, and thus the choice of representation is itself an epistemic decision. Practitioners must continually interrogate the assumptions embedded in their representations, whether human-designed or algorithmically generated, and remain attentive to the ways in which data can mislead if taken at face value.

This interpretive stance becomes even more crucial in the age of frontier technologies. AI systems, for instance, can generate highly polished outputs, elegantly fitted trends, or predictive patterns that appear authoritative. However, the responsibility for evaluating the soundness of these outputs ultimately lies with the human practitioner, not the machine. A model's accuracy, a pattern's statistical reliability, or a prediction's plausibility cannot be accepted merely because they are computationally sophisticated. They must be reasoned through, questioned, and justified with respect to evidence, disciplinary knowledge, and contextual relevance. Thus, reasoning in STEM is inseparable from argumentation. Every interpretation of data is, implicitly or explicitly, an argument, an attempt to connect a claim to evidence through principled reasoning.

Practitioners must articulate not only what the data show, but why their interpretation holds, where uncertainty lies, and how competing explanations compare. In this sense, argumentation becomes the mechanism through which data are transformed into professional judgement. It is the process by which practitioners evaluate the credibility of evidence, adjudicate between competing interpretations, and defend their conclusions within scientific, engineering, and technological communities. For students preparing to enter STEM professions, cultivating this habit of reasoning is foundational: it prepares them for a world in which data are abundant, but sound judgement is scarce.

6.4 Constructing Argumentative Explanations

A scientific or engineering explanation attains its power not through mere description, but through its ability to withstand scrutiny, align with disciplinary principles, and make sense of the evidence in a way that is coherent, defensible, and ethically grounded. In professional contexts, an explanation is persuasive only when it is constructed through a transparent argumentative structure that clarifies the logic behind the practitioner's reasoning. At the core of such an explanation lies the claim, the central statement that articulates what is being asserted about a system, phenomenon, or design. Yet a claim alone is insufficient. The practitioner must provide compelling grounds, whether empirical measurements, computational simulations, observational reports, or statistical analyses. These grounds serve as the evidentiary basis upon which the explanation rests. The warrant, in turn, functions as the intellectual bridge that connects these data to the broader conceptual or theoretical frameworks of the discipline. It communicates why the evidence supports the claim, drawing on established models, principles, or causal mechanisms.

In many cases, stronger explanations require a form of backing, which may include references to canonical literature, validated models, theoretical foundations, or well-established engineering standards. Backing situates the explanation within the intellectual tradition of the field, signalling that the reasoning is not idiosyncratic, but grounded in recognised knowledge. Importantly, rigorous explanations also demonstrate awareness of alternatives. Considering and addressing competing explanations is not a matter of rhetorical strategy but of epistemic responsibility. It demonstrates that the practitioner has scrutinised the full landscape of possibilities and selected the interpretation that best aligns with evidence and disciplinary understanding. Constructing explanations in this argumentative form enhances both communicative clarity and professional credibility (Schwarz et al., 2017). It allows peers, stakeholders, and decision-makers to understand not only what conclusion has been reached, but how it has been reasoned, how uncertainties have been managed, and why it merits acceptance. In an era shaped by frontier technologies, where decisions often carry implications for safety, privacy, equity, and social impact, such clarity is indispensable. For STEM practitioners, this approach strengthens the integrity of their work. For students preparing to enter the profession, it cultivates the intellectual discipline required for responsible, transparent, and ethically informed practice.

6.5 Engaging in Argument from Evidence

As we have seen, argumentation lies at the heart of scientific and engineering progress. It is the means through which the STEM community evaluates the strength of claims, interrogates evidence, and refines collective understanding. Unlike adversarial debate, argumentation in professional STEM practice is a collaborative enterprise that is grounded in shared commitment to accuracy, transparency, and intellectual integrity (Ogunniyi & Iwuanyanwu, 2024). When practitioners engage in argument from evidence, they participate in the ongoing negotiation of knowledge that defines scientific culture. At its core, argumentation requires that claims be presented with clarity and intellectual honesty (Erduran & Park, 2023). Professionals must articulate not only what they conclude but also the limitations of their conclusions, the uncertainties in their measurements, the constraints of their methods, the assumptions built into their computational processes, and the contextual factors that might shape interpretation. This transparency strengthens, rather than weakens, professional authority, for it demonstrates epistemic responsibility.

Engaging in argument from evidence also demands careful evaluation of competing claims, particularly when data appear ambiguous or contradictory. In modern STEM environments, conflicting interpretations often arise from differences in methodological choices, modelling assumptions, or disciplinary perspectives. Thus, argumentation requires practitioners to scrutinise the reasoning that underlies a claim, not merely the numerical outcomes it produces. The choice of sampling strategy, the specification of a model, or the configuration of an AI system can profoundly influence findings. Professionals must therefore question how the evidence was generated before determining what the evidence appears to show. This evaluative process increasingly requires interdisciplinary deliberation. Frontier technologies, such as autonomous systems, AI-enhanced analytics, advanced robotics, and genomic engineering, span multiple domains of expertise. The boundaries between science, engineering, and computation are increasingly porous. Consequently, argumentation within these fields must integrate multiple lenses of reasoning, drawing on expertise that ranges from physics and biology to ethics, human-computer interaction, and data science. Effective practitioners learn to navigate this intellectual complexity through dialogue, critique, and synthesis.

As discussed previously, peer review represents one of the most visible institutional expressions of argumentation from evidence. Whether formal, as in journal review, or informal, as in laboratory meetings, design reviews, or collaborative research discussions, peer critique remains a central mechanism for strengthening the quality and credibility of professional work. Participating in these exchanges requires humility, openness to challenge, and willingness to refine one's claims in response to stronger evidence or more coherent reasoning. Ultimately, engaging in argument from evidence reflects a professional ethos (Simon & Johnson, 2008). It embodies a commitment to truth-seeking, to collective advancement, and to the careful stewardship of knowledge. These commitments are especially important in fields that shape society and the future. For students entering STEM professions, learning to argue from evidence is not merely an academic exercise, it is preparation for responsible participation in the epistemic life of their discipline (Iwuanyanwu, 2020).

6.5.1 A Professional Challenge Task Framing Interdisciplinary Argumentation in Engineering Practice

To deepen our engagement with evidence-based argumentation, we now introduce a high-fidelity engineering challenge that mirrors the epistemic and methodological demands of professional STEM work. This task brings to life the principles we explored earlier, evaluating competing claims, justifying knowledge, and negotiating evidence across disciplines. In contemporary STEM practice, real-world problems rarely fit neatly within a single domain. Challenges such as seismic risk, climate adaptation, infrastructure resilience, and geohazard assessment demand the combined expertise of civil engineers, geologists, and architects, each contributing unique knowledge, assumptions, and methods. By engaging with this task, we step into that complexity, encountering a deliberately ill-structured and epistemically open problem that mirrors the reasoning and decision-making required in research, industry, and policy settings.

Fig. 6.1 Earthquake-induced structural failure scenario

6.5.1.1 The Challenge Scenario

In February 2023, a powerful earthquake struck several inhabited regions of Turkey, causing widespread structural damage. In some areas, entire buildings and heavy structures partially or fully sank into the ground, as if resting on a dense fluid rather than solid earth. This real-world scenario raises urgent technical and epistemic questions: What is happening to the soil? How do seismic forces interact with building foundations? What design or geological factors contribute to such subsidence? Our goal is to investigate these questions and develop a reasoned, evidence-based explanation, defending our argument against competing disciplinary claims. Figure 6.1 shows the aftermath of such an earthquake. Buildings toppled, and in some locations, structures seemed to sink into the earth. Observing these effects, we see that understanding building subsidence requires an interdisciplinary perspective.

Civil engineers remind us that structural integrity and foundation design are critical. Shaking and ground displacement can trigger liquefaction, where saturated soil temporarily behaves like a liquid, causing foundations to settle. Even without liquefaction, intense seismic forces can exert enough pressure to displace foundations. Geologists draw our attention to soil composition and geological conditions. Soft, loose, or sandy soils, weak layers, or proximity to fault lines amplify settlement risks. Even well-constructed buildings may sink if the ground beneath them is unstable. Architects focus on design resilience, incorporating earthquake-resistant materials, flexible structures, and optimized weight distribution. Yet we see that architectural solutions alone are insufficient unless combined with geotechnical and engineering insights.

By bringing these perspectives together, we recognize that no single discipline can address the problem in isolation. We need interdisciplinary collaboration to develop strategies that reduce earthquake risk, strengthen building resilience, and protect communities. With these insights, we are now ready to frame an argumentative approach: How can an earthquake cause a building to sink into the ground? As we tackle this question, we will apply the tools of argumentation, claims, evidence, warrants, counterclaims, and rebuttals, grounded in real-world seismic events. In doing so, we move from theory to practice, stepping fully into the kind of reasoning that shapes professional STEM decision-making.

6.5.1.2 Professional Grouping and Roles

Participants work in multidisciplinary groups representing three professional communities:

(i) Civil Engineering, focusing on foundation behaviour, seismic loads, and structural integrity;
(ii) Geology, analysing soil composition, fault-zone dynamics, and geotechnical properties;
(iii) Architecture, evaluating seismic-resilient design, material selection, and load distribution.

Each group consists of five participants who collaboratively examine the problem, gather and critique evidence, and formulate defensible claims. This structure mirrors interdisciplinary design teams in professional practice.

6.5.1.3 Epistemic Demands of the Task

Because the problem is ill-structured, no single solution can simply be looked up. Instead, practitioners must engage in:

(i) Epistemic cognition, evaluating what constitutes reliable knowledge in their domain;
(ii) Critical comparison of alternative explanations, using disciplinary criteria to adjudicate between competing claims;
(iii) Evidence-based justification, aligning empirical, computational, and conceptual resources with defensible reasoning;
(iv) Judgement and critique, determining which explanations withstand interdisciplinary scrutiny.

6.5.1.4 Gathering and Evaluating Resource Materials

Participants begin by sourcing credible materials, peer-reviewed articles, geotechnical reports, seismic analyses, engineering codes, professional guidelines, or reputable news documentation. These resources must be contextually grounded in the

Table 6.2 Constructing a reasoned explanation

Reporting		
Your challenge	What to do	Your Response
Claims	State your claim either supporting or refuting the factors that lead to the sinking of structures like buildings before, during, and after an earthquake. Keep in mind that the strength of your claim depends on the evidence you provide. Consider the potential arguments or opposing viewpoints that your peers in civil engineering, geology, and/or architecture may put forward before stating your own claim.	
Grounds	State conditions (including evidence, reasons and/or facts) that support your claim(s)	
Source of Evidence	Provide available sources of evidence, including facts, statistics, research findings, expert opinions, or any other relevant information that supports your claim(s).	
Quality of source of evidence	State or cite where you get the data or evidence from, e.g., in terms of quality, you can comment on the reputation of the scientific editor of the source.	
Counterclaims:	Revise your assertions or counterarguments in light of more compelling evidence presented by others.	
Warrant	Explain why the evidence is relevant and how it supports your claims and/or counterclaims.	
Backing	Provide additional evidence or reasoning that supports the warrants.	
Rebuttal	Think of possible weaknesses or flaws in the reasoning of your colleagues from the civil engineering, geology, and architectural groups who are in favour of or against one argument or another and present counter-arguments or alternative perspectives.	

earthquake scenario and must provide data or conceptual tools relevant to soil behaviour, seismic loading, or structural response.

The task requires participants to:

(i) identify the most compelling scientific arguments among their sources;
(ii) analyse the strength of evidence provided;
(iii) evaluate alternative viewpoints and determine whether they become more persuasive when supplemented with additional data.

To support practitioners in approaching this task, participants should gather resource materials from credible sources, such as newspapers, scientific journals, geotechnical reports, or reputable online platforms, that offer conceptual and empirical insights relevant to the earthquake context. They should read and critically examine these materials, identifying which authors or datasets present the most compelling scientific arguments and determining the evidentiary strength of alternative viewpoints. Participants are further encouraged to consider whether competing interpretations might be strengthened if supported by additional data, and to evaluate the scientific merit and credibility of each source. Argumentation should employ the professional test structure *if/and/ then/but /therefore*, supported by calculations, models, geotechnical principles, or structural reasoning as appropriate. When reporting a framework for structuring claims, evidence, warrants, backing, counter-claims, and rebuttals, please refer to Table 6.2. Participants must:

(i) formulate a clear claim;
(ii) present robust grounds and evidence;
(iii) articulate warrants explaining why the evidence supports the claim;
(iv) provide backing from disciplinary theory, standards, or validated models;
(v) consider counterclaims from other disciplinary groups;
(vi) rebut competing interpretations with transparent and defensible reasoning.

6.5.2 Interdisciplinary Perspectives from Expert Groups

The Civil Engineering Group approaches the question by considering the structural integrity of the building. They may argue that during an earthquake, the ground experiences intense shaking and movement, which can cause the soil beneath the building to lose its stability, leading to a phenomenon known as liquefaction. Liquefaction occurs when saturated soil temporarily loses its strength and behaves like a liquid. As a result, the building's foundation becomes unstable, causing it to sink into the ground. Civil engineers further support their claim by pointing out that buildings with shallow foundations are more susceptible to sinking during an earthquake due to the lack of deep anchorage.

The Geology Group may focus on the geological factors that contribute to a building sinking during an earthquake. Geologists could argue that the type of soil and its composition play a crucial role in determining the building's response to seismic activity. For instance, in areas with loose or sandy soil, the ground tends to amplify the shaking caused by an earthquake. This amplification can lead to soil liquefaction, as mentioned earlier. Geologists could highlight the significance of the building's location, noting that if a building is situated on a fault line or near an active seismic zone, the chances of it sinking during an earthquake increase significantly.

The Architectural Group brings a design perspective to the question. Architects could argue that the structural design of a building can influence its response to seismic activity. They can also emphasize the importance of incorporating earthquake-resistant features into the building's design, such as flexible materials, reinforced concrete, and steel frames. By implementing these measures, architects aim to minimize the building's vulnerability to sinking during an earthquake. Additionally, architects consider the height and weight distribution of the building. Tall and heavy structures are more prone to sinking as they exert greater pressure on the ground, especially if the foundation is not adequately designed to withstand seismic forces. These examples demonstrate not only the content knowledge of each field but the argumentative practices that distinguish professional reasoning, thus, demonstrating how claims are constructed, how evidence is mobilised, and how assumptions are made explicit.

6.5.3 Communicating Findings

In professional settings, high-stakes decisions require transparent communication among engineers, geoscientists, architects, regulators, and public stakeholders. Participants therefore communicate their findings through a structured report, oral presentation, or formal design review. This step ensures that:

(i) reasoning is explicit and publicly defensible;
(ii) uncertainty is made visible rather than concealed;
(iii) methodological decisions are justified;
(iv) interdisciplinary critiques can be integrated into final recommendations.

6.5.4 Insights from Earthquake-Induced Structural Failure Scenario

To understand how an earthquake can cause a building to sink into the ground, it is helpful to consider perspectives from civil engineering, geology, and architecture. Each discipline provides unique insights into the phenomenon, highlighting the interplay between soil behaviour, structural design, and seismic forces. Below, we summarize the key arguments, counterclaims, rebuttals, and surrebuttals from each perspective (Table 6.3).

Table 6.3 Sample of possible arguments and counterarguments by expert groups

Reporting	
Participants	Argumentative Discourse
Civil Engineering Argument	• If an earthquake generates seismic waves, then the ground beneath a building can experience shaking and displacement. • And if the soil undergoes liquefaction due to the seismic activity, then its strength decreases, potentially leading to the sinking of a building's foundation. • Therefore, civil engineers assert that the dynamic nature of the soil during an earthquake can contribute to the settlement and sinking of buildings.
Geology Counterargument	• If not that all soil types are prone to liquefaction, then it is essential to consider the geological characteristics of the site. • And if the geological composition of the soil is stable and resistant to liquefaction, then the building may not sink into the ground solely due to seismic activity. • Therefore, geologists argue that the geological factors play a crucial role in determining the susceptibility of a building to sink during an earthquake.

(continued)

Table 6.3 (continued)

Reporting	
Participants	Argumentative Discourse
Civil Engineering Rebuttal	• If seismic waves exert significant pressure on the foundation of a building, then even stable soil can undergo settlement and sinking. • But if the ground experiences intense shaking and displacement during an earthquake, then the building's foundation may still be at risk of sinking, irrespective of the soil type. • Therefore, civil engineers emphasize that while geological conditions are important, the seismic forces acting on the building during an earthquake can also contribute to its sinking.
Architecture Argument	• If a building is not structurally reinforced to withstand seismic forces, then it may succumb to settlement and sinking during an earthquake. • And if the design and construction quality of the building do not consider seismic resilience, then the structure is more vulnerable to sinking. • Therefore, architects assert that incorporating seismic-resistant design principles is essential to prevent buildings from sinking into the ground during earthquakes.
Civil Engineering Counterargument	• If the foundation of a building lacks proper support and reinforcement, then it may be susceptible to sinking during an earthquake. • And if the interaction between the structure and the ground is not adequately addressed, then even well-designed buildings can experience settlement. • Therefore, civil engineers argue that structural integrity and foundation stability are crucial factors in preventing buildings from sinking into the ground.
Architecture Rebuttal	• If architects integrate advanced seismic-resistant technologies and design strategies, then they can enhance a building's ability to withstand seismic forces. • But if the structural collaboration between architects and civil engineers is lacking, then the building may still be at risk of sinking during an earthquake. • Therefore, architects emphasize the importance of interdisciplinary cooperation to address the complex interactions between building design, soil dynamics, and seismic activity.
Closing Thoughts	The report discusses the importance of considering various factors in building stability, including dynamic soil behavior, seismic forces, geological factors, and seismic-resistant design. By addressing both structural and geological considerations, professionals can develop holistic solutions to mitigate the risk of building sinking and enhance overall structural resilience during seismic events.

6.5.5 Civil Engineering Perspective

Open Remark: Civil engineers play a crucial role in designing structures that withstand natural disasters, including earthquakes. They examine factors such as soil dynamics, foundation design, and structural integrity to understand why buildings may sink during seismic events.

Supporting Point: During an earthquake, seismic waves cause the ground to shake and shift. This movement can trigger liquefaction, where saturated soil temporarily loses its strength and behaves like a liquid, potentially causing the building's foundation to sink.

Counterclaim (Geology): Geologists may argue that not all soil types are prone to liquefaction. The geological composition of a site strongly influences whether a building will sink.

Rebuttal (Civil Engineering): Civil engineers emphasize that even if liquefaction is not the main cause, intense shaking and ground displacement can still exert significant pressure on a building's foundation, leading to settlement and potential sinking.

Surrebuttal (Geology): Geologists acknowledge the role of seismic waves but note that site-specific conditions, such as weak or compressible soil layers, can worsen the sinking of a building.

6.5.6 Geological Perspective

Open Remark: Geologists study the Earth's materials, structure, and processes, providing insights into how soil and geological factors affect buildings during earthquakes.

Supporting Point: Soil composition and characteristics are critical. Loose, soft, or sandy soils are more likely to settle and contribute to a building sinking when subjected to seismic forces.

Counterclaim (Architecture): Architects may argue that the design and quality of a building can prevent sinking, even on unstable soils, emphasizing structural measures and reinforcements.

Rebuttal (Geology): While building design matters, geological conditions ultimately determine ground movement. Even well-constructed structures can settle or sink if the soil is weak.

Surrebuttal (Architecture): Architects stress incorporating geotechnical knowledge into the design process. Proper foundation design, soil stabilization, and seismic-resistant strategies can mitigate the risk of sinking.

6.5.7 Architectural Perspective

Open Remark: Architects focus on designing buildings that are functional, safe, and aesthetically pleasing. They address seismic risks through earthquake-resistant design principles.

Supporting Point: Properly designed buildings can resist lateral forces during earthquakes. However, inadequate foundation support or insufficient reinforcement can lead to settlement and sinking.

Counterclaim (Civil Engineering): Civil engineers highlight that the interaction between the structure and the ground is critical. Even well-designed buildings can sink without a solid foundation and structural support.

Rebuttal (Architecture): Architects point to innovative materials, structural systems, and seismic-resistant design techniques that enhance a building's resilience and minimize the risk of sinking.

Surrebuttal (Civil Engineering): Collaboration between architects and geotechnical engineers is essential to address interactions between building design, soil behaviour, and seismic forces. Interdisciplinary cooperation ensures holistic solutions and enhances structural resilience.

6.5.8 Summary of the Challenge's Contribution to Professional Practice

The interdisciplinary perspectives of civil engineering, geology, and architecture provide valuable insights into why buildings may sink during earthquakes. By considering the dynamic interactions between soil properties, structural design, and seismic forces, professionals can develop comprehensive strategies to improve building resilience. Collaboration across disciplines is essential for mitigating earthquake risks and ensuring the safety and stability of the built environment. This challenge integrates the intellectual, methodological, and ethical commitments essential to STEM professionalism. It demands not only technical expertise but disciplined reasoning, argumentative clarity, and interdisciplinary sensitivity. Engaging with complex real-world problems in this manner prepares emerging professionals to navigate uncertainty, evaluate evidence responsibly, and contribute to informed technological and societal decision-making.

6.6 Modelling and Applying Scientific Knowledge

Modelling occupies a central position in STEM practice. Whether conceptual, mathematical, physical, or computational, models serve as the intellectual scaffolds through which practitioners make sense of complex systems. They allow professionals to represent phenomena that cannot be directly observed, to test hypotheses under controlled conditions, to explore counterfactual scenarios, and to generate predictions that inform design, policy, and technological development. In professional settings, modelling is more than a technical procedure, it is an epistemic activity. When practitioners construct a model, they exercise judgement about what features of the world to include, what to simplify, and what assumptions to adopt. These decisions shape how the model behaves, what it illuminates, and what it inevitably obscures. Thus, modelling is simultaneously an act of abstraction and an act of argumentation.

A model asserts that certain aspects of reality are relevant, others negligible, and still others beyond the scope of representation. Practitioners apply models to test hypotheses, particularly when empirical investigation is impractical, dangerous, or ethically constrained. Computational simulations, for example, allow engineers to explore structural stresses without building prototypes; climate scientists to forecast environmental change; and biomedical researchers to examine biological processes at scales impossible to observe directly. Models provide a means of anticipating consequences, evaluating alternatives, and informing evidence-based decision-making across STEM domains.

Yet because models are simplifications, they carry inherent limitations. They are not replicas of the world but approximations designed for specific purposes. Their assumptions, about scale, behaviour, boundary conditions, or interactions, must be made explicit if the model is to be used responsibly. Practitioners therefore bear a crucial ethical responsibility: to ensure that models are not misinterpreted, overextended, or applied outside their valid domain. This responsibility becomes especially significant in high-stakes contexts such as environmental forecasting, public health, automated decision systems, or engineering safety assessments, where stakeholders may depend on model outputs without fully understanding their constraints. Responsible modelling requires practitioners to reflect on the consequences of their methodological choices, to evaluate the robustness of their predictions, and to communicate uncertainty with precision and clarity.

In an era shaped by AI and other frontier technologies, models increasingly take the form of machine-learning systems whose internal mechanisms may be opaque even to their designers. This heightens the need for rigorous interrogation, interpretability, and ethical oversight. For emerging STEM professionals, the practice of modelling fosters intellectual discipline: it teaches how to simplify without distorting, how to reason about systems that cannot be directly observed, and how to connect theoretical understanding with practical application. For experienced practitioners, modelling remains indispensable to innovation, enabling the continual refinement of knowledge and the development of solutions that respond to evolving technological and societal demands.

6.7 Evaluating Scientific Claims

Evaluating scientific claims is a disciplined, deliberative process that demands both critical literacy and mature professional judgement. In contemporary STEM environments, where data are plentiful, models are complex, and automated systems frequently generate outputs that appear authoritative, evaluation has become as important as discovery itself. The capacity to interrogate claims is therefore a defining feature of responsible STEM practice. This evaluative work begins with examining the quality, provenance, and integrity of the evidence on which a claim is built. Practitioners must determine whether the data were generated through reliable instruments, valid sampling strategies, and coherent methodological procedures. This requires attention not only to the techniques employed but also to the assumptions underpinning them.

A claim grounded in flawed measurement or incomplete sampling can only yield fragile conclusions, no matter how sophisticated the analytical tools. Judging methodological robustness extends further into assessing sources of uncertainty, limitations in design, and the reliability of computational or statistical techniques. The emergence of automated analysis, especially AI-driven pattern recognition, has added new dimensions to this scrutiny. Practitioners must now evaluate whether algorithmic biases, model mis-specification, or training-data limitations have influenced the outputs they receive. This responsibility cannot be delegated to machines; the burden of interpretation ultimately rests with the human professional.

Critically evaluating claims also requires distinguishing between descriptive associations and causal explanations. Correlations, no matter how striking, do not establish underlying mechanisms. STEM practitioners must continually ask whether the evidence supports causal reasoning or merely reveals superficial patterns. This distinction becomes particularly significant in domains such as biomedicine, environmental modelling, and socio-technical systems, where policy decisions may rely on accurate causal understanding. Replicability and generalisability represent additional pillars of claim evaluation. Practitioners must consider whether findings can be reproduced under comparable conditions and whether they apply across contexts, populations, or environmental conditions. In frontier technological fields, such as AI, CRISPR-based gene editing, or IoT-driven analytics, context often shapes outcomes in ways that are subtle but profound. Evaluating claims therefore means assessing both their internal coherence and their external reach. Finally, evaluation is never purely technical; it is always shaped by ethical, contextual, and societal considerations. Claims do not exist in a vacuum. They influence communities, ecosystems, and public trust. A STEM professional must therefore balance healthy skepticism with intellectual openness, recognising when a claim is sufficiently supported while remaining vigilant toward overgeneralisation, misrepresentation, or unwarranted certainty. This balanced stance, critical yet constructive, is essential for maintaining the integrity of professional STEM judgement.

6.8 Ethical Reasoning in Scientific Practice

Ethical reasoning is inseparable from scientific and technological work. Every decision a STEM professional makes, whether designing a study, interpreting data, building an algorithm, modifying a biological system, or advising on policy, carries social, environmental, and moral consequences. In an era shaped by frontier technologies, where the pace of innovation often outstrips the development of regulatory frameworks, ethical reasoning is no longer peripheral to scientific competence; it is foundational. Responsible ethical practice begins with the stewardship of data. As practitioners increasingly work with sensitive, high-volume, or AI-processed datasets, the obligation to protect privacy, ensure fairness, and maintain security becomes paramount. Ethical reasoning requires professionals to consider not only the legality of their data practices but also the broader implications for dignity, autonomy, and trust. Decisions about how data are collected, stored, shared, and analysed must be guided by principles that prioritise the well-being of individuals and communities. Transparency represents another essential dimension of ethical STEM work. Practitioners must report methods, limitations, uncertainties, and interpretations with honesty and clarity. This includes resisting pressures, whether institutional, commercial, or political, to exaggerate findings, emphasise favourable results, or obscure methodological weaknesses. Transparency fosters credibility and ensures that scientific knowledge contributes to society in ways that are aligned with truth and justice.

Ethical scientific practice also requires awareness of the societal and environmental implications of technological innovation. Engineers and technologists must consider how their creations reshape labour markets, redistribute power, influence ecological systems, and affect the lives of vulnerable populations. Researchers working with emerging technologies such as AI or gene-editing tools must anticipate the risks of technological displacement, unintended consequences, or widening inequities. Ethical reasoning demands that practitioners evaluate not only what can be done, but what should be done, and for whose benefit. Recognition of power dynamics becomes critical when research involves communities, interdisciplinary teams, or cross-sector collaboration.

Ethical engagement requires sensitivity to differing worldviews, equitable participation, and respect for local knowledge systems. Practitioners must consider how their authority, expertise, or institutional position influences decisions and whose voices may be marginalized in the process. Above all, ethical reasoning in STEM is anchored in professional integrity. It calls for a steadfast commitment to honesty, accountability, and the public good. In the age of frontier technologies, where AI systems act on behalf of humans, biotechnologies can alter life at its foundations, and robotics increasingly shape human environments, ethical judgement becomes a defining attribute of professional excellence. It ensures that innovation advances human flourishing rather than undermining it. For practitioners and aspiring STEM professionals alike, ethical reasoning is not an abstract ideal. It is a daily practice, a moral orientation, and a habit of mind that must inform every stage of scientific inquiry and technological development.

6.9 Reflective Considerations for Practitioners

At this stage of our inquiry, we invite you into a deeper reflection to strengthen our engagement with the ways disciplinary cultures, methodological choices, and ethical commitments shape both the knowledge we produce and the decisions we make. Rather than offering fixed answers, the questions raised here aim to encourage us to explore and interrogate our practices. You might begin by reflecting on the epistemic norms that govern your field. Every discipline, whether physics, engineering, biotechnology, or computational science, has implicit standards for what counts as credible evidence. How do these norms guide your own evaluations, and in what ways might they obscure alternative forms of knowledge or more diverse ways of knowing? Consider also the interpretive responsibilities you assume when engaging with data. Whether the outputs originate from human analysis, automated pipelines, or AI-assisted systems, each dataset carries assumptions, limitations, and potential sources of bias. What strategies help you recognise fragile inferences, identify missing context, or detect unwarranted certainty in the patterns you observe?

You may find it valuable to examine how your investigative designs shape the argumentative richness of your practice. Do your projects create opportunities for colleagues or students to reason with evidence, explore competing explanations, and articulate defensible claims? How might you refine the structure of your inquiries to better support these forms of collaborative knowledge-building? Modelling too invites reflection. Every model, conceptual or computational, represents a choice about what to foreground and what to omit. How do these choices influence the explanations you formulate, the decisions you endorse, and the trust that stakeholders place in your conclusions? Finally, ethical tensions are an inevitable part of scientific and technological work, particularly in fields shaped by rapid innovation. As you navigate dilemmas related to data stewardship, environmental consequences, technological risk, equity, or public accountability, how might your scientific reasoning provide clarity, humility, and moral direction? These reflective provocations are not tasks to complete, but enduring questions that accompany the daily work of the STEM professional. By engaging with them, practitioners and aspiring practitioners alike strengthen the intellectual and ethical foundations of their practice, a foundation that will be increasingly vital in the decades ahead.

6.10 Chapter Summary

This chapter has reconceptualised scientific practices as a dynamic, interconnected system that sustains rigorous STEM reasoning, responsible decision-making, and ethically grounded professional judgement. We have emphasised that designing investigations, interpreting data, constructing explanations, modelling phenomena, evaluating claims, and engaging in argument from evidence are not isolated technical tasks. Rather, they form an ecosystem of cognitive and ethical actions that define the work of contemporary STEM practitioners. Within this framework, scientific practice becomes the intellectual architecture that enables professionals to engage confidently and critically with complex problems. It provides the means to navigate uncertainty, to interrogate data, whether human-generated or produced through automated systems, and to articulate explanations that withstand scrutiny. These practices cultivate the habits of mind that allow practitioners to interrogate assumptions, recognise limitations, and situate their decisions within broader societal and environmental contexts.

In an age defined by frontier technologies, where AI, biotechnology, computational modelling, and interconnected digital systems increasingly shape how knowledge is produced and applied, the importance of these scientific practices cannot be overstated. They safeguard the integrity of inquiry, ensuring that STEM work remains grounded in evidence, reason, and ethical responsibility. For both current practitioners and the emerging professionals who will soon inherit these responsibilities, robust scientific practice is not merely a methodological competency, it is a form of intellectual citizenship. It positions STEM work as a human-centred endeavour, capable of advancing innovation while honouring the moral commitments that underpin the pursuit of knowledge.

6.10.1 Looking Ahead to Chap. 7

In Chap. 6, we explored how designing investigations, interpreting data, constructing explanations, modelling phenomena, and engaging in argument from evidence collectively enable practitioners to navigate complexity, interrogate assumptions, and situate decisions within broader societal and environmental contexts. These practices provide the cognitive and ethical foundation for professional STEM work, preparing us not only to reason critically but also to act responsibly when confronting uncertainty. Having examined how scientific practices form a dynamic, interconnected system that underpins rigorous STEM reasoning, responsible decision-making, and ethically grounded professional judgment, we now turn our attention to how these practices intersect with engineering design thinking.

Chapter 7 builds directly on this foundation by examining engineering design thinking as a structured yet creative approach to problem-solving in STEM contexts. We will explore how practitioners progress from problem identification to idea generation, conceptual design, prototyping, testing, and iterative refinement, all while integrating argumentation, stakeholder perspectives, and ethical reasoning. Creativity emerges as a central component, enabling us to expand solution spaces, imagine alternative approaches, and enhance innovation, while ethical reflection guides our decisions around safety, sustainability, and social responsibility.

By linking scientific practice with engineering design thinking, Chap. 7 illuminates the ways in which rigorous cognition, creative ideation, and principled judgment converge in professional STEM work. As we engage with this chapter, we will see how the habits of mind cultivated through scientific practice are essential not only for evaluating knowledge but also for generating and defending solutions that are technically sound, ethically responsible, and socially meaningful. In this way, Chaps. 6 and 7 form a continuum, moving from understanding how knowledge is constructed and interrogated to learning how that knowledge can be applied creatively, collaboratively, and responsibly to address complex, real-world challenges.

6.11 Reflection Questions

1. How do you determine what counts as credible evidence in your field, and how have these standards evolved in response to emerging technologies such as AI, advanced analytics, or automated measurement tools?
2. In your investigative work, how do methodological choices, sampling, instrumentation, modelling strategies, computational pipelines, shape the kinds of explanations and arguments you can ultimately construct?
3. When interpreting data, what practices help you distinguish between meaningful patterns and artefacts generated by algorithms, noise, or hidden assumptions within your analytical tools?
4. How do you evaluate the reliability of AI-generated outputs or computational predictions, and what processes do you use to integrate human judgement into automated decision-making environments?
5. In what ways do you create opportunities within your professional context (or classroom) for colleagues or students to engage in evidence-based argumentation, critique assumptions, and examine competing explanations?
6. How do the modelling choices you make, conceptual, mathematical, physical, or computational, shape the interpretations, decisions, or policy recommendations that arise from your work?
7. What strategies do you use to navigate uncertainty, especially in contexts where data are incomplete, ambiguous, or potentially biased? How do these strategies influence your argumentation and decision-making?
8. Where do ethical tensions emerge in your practice, such as data privacy, equity, environmental responsibility, or algorithmic bias, and how might scientific reasoning help you address these tensions with integrity?

9. How do you balance the need for innovation with the responsibility to anticipate and mitigate unintended consequences, particularly in work involving frontier technologies like AI, gene editing, IoT, or robotics?
10. In your experience, how does collaborative argumentation, peer review, interdisciplinary dialogue, or group critique, strengthen the rigour, transparency, and societal relevance of STEM decision-making?

References

Akiri, E., Tal, M., Peretz, R., Dori, D., & Dori, Y. J. (2020). STEM graduate students' systems thinking, modeling and scientific understanding—The case of food production. *Applied Sciences, 10*(21), 7417. https://doi.org/10.3390/app10217417

Dym, C. L., Agogino, A. M., Eris, O., Frey, D. D., & Leifer, L. J. (2005). Engineering design thinking, teaching, and learning. *Journal of Engineering Education, 94*(1), 103–120. https://doi.org/10.1002/j.2168-9830.2005.tb00832.x

Erduran, S., & Park, W. (2023). Argumentation in physics education research: Recent trends and key themes. In M. F. Taşar & P. R. L. Heron (Eds.), The international handbook of physics education research: Learning physics (pp. 16–32). AIP Publishing.

Iwuanyanwu, P. N. (2020). Nature of problem-solving skills for 21st century STEM learners: What teachers need to know. *Journal of STEM Teacher Education, 55*(1), 27–40. https://doi.org/10.30707/JSTE55.1/MMDZ8325

Lee, H.-Y., Wu, T.-T., Lin, C.-J., Wang, W.-S., & Huang, Y.-M. (2023). Integrating computational thinking into scaffolding learning: An innovative approach to enhance science, technology, engineering, and mathematics hands-on learning. *Journal of Educational Computing Research, 62*(2), 211–247. https://doi.org/10.1177/07356331231211916

Lockwood, E., & Mørken, K. (2021). A call for research that explores relationships between computing and mathematical thinking and activity in RUME. *International Journal of Research in Undergraduate Mathematics Education, 7*, 404–416. https://doi.org/10.1007/s40753-020-00129-2

Mafokwane, M. P. (2026). Modelling teachers' self-directed learning and engagement in a STEAM-based professional development program: A structural equation modelling study in South Africa. *EURASIA Journal of Mathematics, Science and Technology Education, 22*(3), em2788. https://doi.org/10.29333/ejmste/17971

Ogunniyi, M., & Iwuanyanwu, P. N. (2024). Analysis of teachers' perspectives towards the use of IKS to improve STEM education for sustainable development. *African Journal of Research in Mathematics, Science and Technology Education, 28*(3), 319–329. https://doi.org/10.1080/18117295.2024.2352980

Osborne, J. F. (2019). Not "hands on" but "minds on": A response to Furtak and Penuel. *Science Education, 103*, 1280–1283. https://doi.org/10.1002/sce.21543

Peretz, R., Levi-Soskin, N., Dori, D., & Dori, Y. J. (2024). Assessing engineering students' systems thinking and modeling based on their online learning. *IEEE Transactions on Education, 67*(3), 483–490. https://doi.org/10.1109/te.2024.3379218

Popper, K. R. (2002). *Conjectures and refutations: The growth of scientific knowledge*. Routledge.

Schwarz, B. B., Rapanta, C., & Asterhan, C. S. (2017). Argumentation in education. In A. A. Ioannou, S. Papadopoulou, & T. Mikropoulos (Eds.), *Handbook of educational technology research* (pp. 287–302). Springer. https://doi.org/10.1007/978-3-319-60013-0_20

Simon, S., & Johnson, S. (2008). Professional learning portfolios for argumentation in school science. *International Journal of Science Education, 30*(12), 1701–1721. https://doi.org/10.1080/09500690701579572

Chapter 7
Engineering Design Thinking, Creativity, and Ethical Reasoning

7.1 Introduction

Engineering design thinking provides a unifying professional logic for navigating uncertainty, complexity, and responsibility in contemporary STEM practice. In this chapter, we position engineering design thinking as a disciplined yet creative mode of professional reasoning through which practitioners and emerging professionals frame problems, generate and evaluate solutions, and justify decisions in contexts shaped by rapid technological change and profound socio-ethical consequence (Cross, 2008; Dym et al., 2005). We demonstrate how creativity expands the design space through modelling, prototyping, and iterative refinement, while ethical reasoning guides judgments about safety, sustainability, equity, and societal impact (Baldassarre et al., 2024; Van de Poel et al., 2018). Argumentation is foregrounded as the epistemic and communicative practice that enables design teams to evaluate alternatives, defend design decisions, and engage responsibly with diverse stakeholders (Erduran et al., 2015; Lee et al., 2024).

Across STEM professions, we increasingly confront problems that resist neat definition and linear solution. Contemporary engineering challenges, ranging from artificial intelligence governance and renewable energy transitions to biomedical innovation and infrastructure resilience, are characterized by uncertainty, competing constraints, and ethical consequence. In such contexts, technical proficiency alone is insufficient. What distinguishes effective professional practice is the capacity to exercise judgment and to reason with evidence, imagine alternatives, weigh values, and defend decisions publicly and transparently (Simon, 1996; National Academy of Engineering [NAE], 2020). In this chapter, we engage with engineering design thinking not as a pedagogical heuristic for students alone, but as a professional epistemology that underpins responsible STEM practice. Engineering design thinking has long been recognized as a distinctive form of problem solving that integrates analytical reasoning, creativity, and contextual sensitivity (Dym et al.,

P. N. Iwuanyanwu, *Empowering STEM Thinkers Through Argumentation*, Contemporary Trends and Issues in Science Education 69,
https://doi.org/10.1007/978-3-032-27906-4_7

2005). This chapter's contribution extends this view by foregrounding argumentation and ethical reasoning as central, rather than peripheral, to design practice. Engineering design, as presented throughout this chapter, is not merely the production of artifacts; it is the disciplined justification of decisions at the intersection of evidence, values, and public responsibility (Winner, 1986).

Figure 7.1 captures the chapter's central argument that engineering design thinking is a cyclical, judgment-oriented, and ethically accountable mode of professional reasoning rather than a linear sequence of technical steps. The diagram's circular structure foregrounds iteration, emphasising that design unfolds through continuous movement among interconnected phases rather than toward a single, final solution. At the centre is STEM Learning & Practice, highlighting that design thinking is not an abstract method but a core mode of reasoning that shapes how STEM practitioners act, decide, and justify their work in real-world contexts. Encircling this core are five interrelated phases, problem scoping, concept generation, modelling and prototyping, testing and decision-making, and ethical and argumentative review, which together reinforce the non-linear and recursive nature of engineering design. This structure aligns with the chapter's portrayal of contemporary STEM problems as ill-structured, uncertain, and value-laden, requiring practitioners to revisit earlier stages as new evidence, constraints, or ethical considerations emerge.

Concept generation and modelling and prototyping represent the creative expansion of the design space. These phases reflect how imagination, creativity, and representational tools enable designers to explore alternatives, make ideas tangible, and iteratively refine solutions. The framework reinforces the claim that creativity is not opposed to rigor, but is disciplined through modelling, evidence, and systematic evaluation. Testing and decision-making highlights the evaluative dimension of design thinking, where empirical evidence, performance criteria, risk considerations, and stakeholder implications are weighed. As discussed in Chaps. 3 and 4, decisions are portrayed not as purely technical optimisations but as reasoned judgments that must be justified in light of uncertainty, trade-offs, and consequences.

Crucially, ethical reasoning and argumentation permeate all stages of the cycle rather than appearing as a terminal or add-on component. This visual design directly supports the chapter's argument that argumentation functions as the epistemic and communicative practice through which design teams evaluate alternatives, defend decisions, and engage responsibly with diverse publics. In this sense, ethical considerations, such as safety, sustainability, equity, and societal impact are presented as integral to problem framing, solution generation, testing, and justification alike. This emphasis is visually reinforced by the outer ring and directional arrows associated with ethical reasoning and argumentation, which encircle and connect each phase. Ethical judgment, in this sense, informs how problems are framed, which concepts are pursued, how models are interpreted, and which design decisions are ultimately legitimised. By integrating creativity (through concept generation and prototyping), analytical reasoning (through modelling and testing), and normative judgment (through ethical–argumentative review), Fig. 7.1 captures engineering design thinking as a professional epistemology rather than a narrow technical procedure.

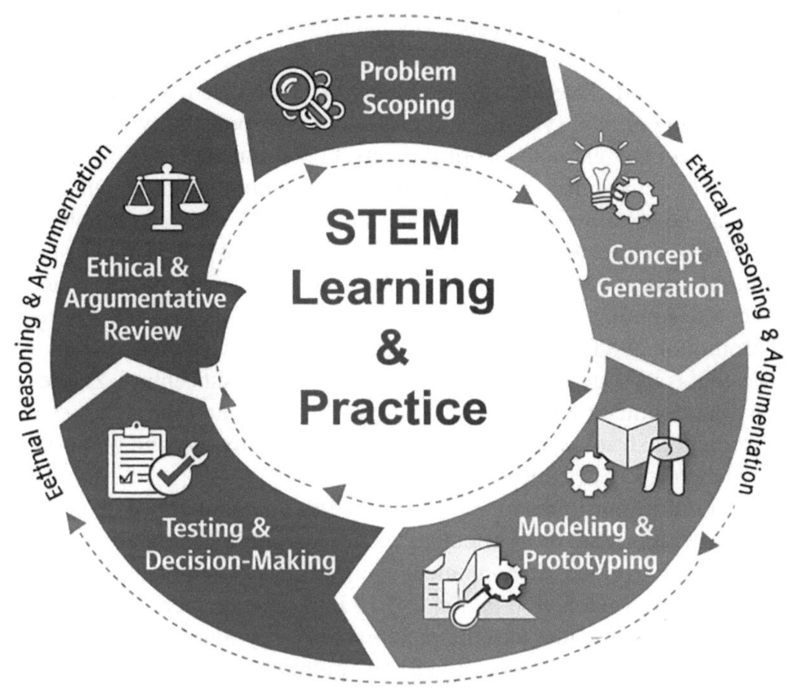

Fig. 7.1 Integrated engineering design thinking framework

7.2 Engineering Design and Problem Scoping in Frontier Technologies

Professional engineering practice begins not with solutions, but with how problems are framed. In frontier and emerging technologies, such as artificial intelligence, biotechnology, advanced materials, energy systems, and cyber-physical infrastructures, problem scoping is a decisive act that shapes downstream decisions and societal consequences (Van de Poel et al., 2018). We emphasize that problem scoping is inherently an argumentative and ethical practice. Engineers advance claims about what the problem is, justify these claims using technical data, stakeholder requirements, regulatory standards, and contextual knowledge, and anticipate counterclaims arising from alternative framings (Erduran et al., 2015). For example, framing algorithmic systems primarily as optimisation problems can obscure questions of bias, accountability, and social harm; similarly, defining energy challenges solely in terms of efficiency can marginalise issues of environmental justice and access (Martin et al., 2021).

Effective problem scoping in STEM professional contexts therefore requires surfacing underlying assumptions and value commitments, integrating technical and social constraints, and explicitly engaging socio-STEM and socio-scientific dimensions. Socio-scientific and socio-technical issues (SSI/STI) are not peripheral concerns appended after technical decisions are made. They are constitutive of the design context itself. Engineering problems are embedded within communities, ecosystems, and political systems, and professional responsibility demands that we recognize these entanglements from the outset. Without such engagement, STEM education risks reinforcing a widely critiqued assumption that expanding STEM initiatives alone is sufficient to cultivate an informed and responsible citizenry (Zeidler, 2014). Zeidler argues that when STEM education is detached from broader sociocultural contexts, it fails to address the complex ethical, political, and global challenges confronting contemporary societies. From this perspective, the widespread promotion of STEM as a universal solution reflects a deficit-oriented view of education that is ill-suited to the demands of the twenty-first century. Zeidler further contends that dominant STEM practices must be reoriented through sociocultural and socioscientific frameworks if they are to meaningfully respond to the escalating challenges facing STEM education.

Research on responsible and value-sensitive design similarly demonstrates that ethical considerations are most effective when incorporated early in the design process rather than retrofitted after technical decisions have been finalised (Baldassarre et al., 2024; Spiekermann & Winkler, 2020). Early engagement with socio-STEM issues enables design teams to anticipate tensions, negotiate trade-offs, and avoid narrow solutionism. From a professional standpoint, problem scoping is also a communicative act. Engineers must articulate the rationale for their framing to colleagues, managers, clients, regulators, and the public. Argumentation here functions to justify why certain constraints are prioritized, why specific stakeholders are consulted, and why alternative framings are set aside. In this sense, problem scoping establishes the moral and epistemic boundaries within which subsequent design decisions unfold. Table 7.1 illustrates how problem scoping shifts when engineering challenges are framed as socio-technical rather than purely technical problems. The table aligns well with Fig. 7.1, reinforcing the argument that ethical reasoning and argumentation are foundational to problem scoping.

As can be seen in Table 7.1, in both artificial intelligence and energy systems, technical framings prioritise optimisation, efficiency, and performance, often limiting ethical consideration to downstream evaluation or regulatory compliance. By contrast, socio-technical scoping integrates ethical reasoning into the initial definition of the problem, broadens the range of stakeholders considered, and foregrounds long-term societal and environmental consequences. This contrast reinforces the chapter's argument that responsible engineering design depends not only on technical proficiency but on the capacity to frame problems in ways that make values, trade-offs, and public accountability explicit from the outset.

Table 7.1 Contrasting narrow technical problem framing and socio-technical problem scoping

Frontier technology domain	Narrow technical problem framing	Socio-technical problem scoping
Artificial Intelligence (AI)	Assumes the problem is primarily one of optimisation, accuracy, or computational efficiency. Focuses on improving algorithms, performance metrics, and system scalability.	Frames AI as a socio-technical system embedded in social, legal, and cultural contexts. Considers bias, transparency, accountability, governance, and societal impact alongside technical performance.
Biotechnology & Genetic Engineering	Treats challenges as technical issues of precision, efficacy, and biological control. Emphasises laboratory success and functional outcomes.	Recognises ethical concerns related to consent, equity, long-term ecological effects, and moral boundaries of intervention in life processes. Engages public values, regulation, and social trust.
Energy systems & transitions	Defines problems in terms of efficiency, output, cost reduction, and technological feasibility. Prioritises system performance and optimisation.	Expands the scope to include environmental justice, access, community impact, sustainability, and intergenerational equity. Integrates social acceptance and policy considerations.
Advanced materials & nanotechnology	Focuses on material properties, performance enhancement, and manufacturability. Assumes risks are manageable through technical controls.	Considers health, environmental, and lifecycle implications, including exposure risks, waste, and long-term societal consequences of deployment at scale.
Cyber-physical systems & smart infrastructure	Frames problems as issues of reliability, automation, and system integration. Emphasises efficiency, control, and resilience from a technical standpoint.	Addresses data privacy, surveillance, security vulnerabilities, governance, and impacts on public autonomy, labour, and social equity.
Underlying assumptions	Technology is value-neutral; social issues can be addressed after technical optimisation.	Technology is value-laden; ethical and social considerations are integral to problem definition from the outset.
Stakeholders considered	Primarily engineers, designers, clients, and technical experts.	Includes diverse stakeholders such as users, affected communities, policymakers, regulators, and the public.
Ethical visibility	Ethics treated as an add-on or compliance requirement after design decisions are made.	Ethics embedded throughout problem framing, design choices, testing, and justification.
Long-term consequences	Often underexamined or deferred; emphasis on short-term functionality and performance.	Explicitly anticipates long-term societal, environmental, and intergenerational consequences.

7.3 Argumentative Reasoning in the Concept Development Stage of Engineering Design

Once a problem has been scoped, engineering design proceeds through the generation and evaluation of alternative concepts. At this stage, argumentation becomes central to professional reasoning. Concept development is not merely ideation; it is a process of disciplined comparison in which competing possibilities are advanced, scrutinized, and justified (Lee et al., 2024). In professional design teams, argumentative reasoning is collaborative rather than adversarial. Practitioners articulate claims about why one concept is preferable to another, support these claims with simulation results, empirical data, feasibility analyses, and cost models, and connect evidence to claims through engineering principles, standards, and precedents (Cross, 2008; Dym et al., 2005). Importantly, argumentation also involves recognizing uncertainty and limitation. Claims are typically qualified rather than absolute, reflecting the provisional nature of design knowledge (Iwuanyanwu, 2020). Equally important is the capacity to anticipate rebuttals, such as failure modes, ethical risks, or unintended consequences, and to respond transparently.

A Toulmin-inspired argumentative structure often operates implicitly in professional design reviews, linking claims, data, warrants, qualifiers, and rebuttals in ways that support reasoned judgment under uncertainty (Erduran et al., 2015; Ogunniyi, 2007). The argumentative structure may be presented in the design review as follows:

(i) Claim: This design concept best satisfies the defined requirements.
(ii) Data: Modelling results, experimental tests, benchmarking studies.
(iii) Warrant: Engineering principles or standards linking evidence to the claim.
(iv) Qualifier: Conditions and assumptions under which the claim holds.
(v) Rebuttal: Anticipated limitations, risks, or ethical concerns.

Argumentation thus functions as a quality-control mechanism, ensuring that creativity remains accountable to evidence, professional norms, and societal responsibility. For example, as discussed in Chap. 5, Sect. 5.4, a design team may claim that a particular material choice optimizes the strength-to-weight ratio, support this claim with modelling data, justify it through established materials science principles, and qualify it by acknowledging performance degradation under extreme conditions. Equally important is the capacity to anticipate rebuttals, such as failure modes, ethical risks, or unintended consequences, and to respond transparently. Argumentation therefore disciplines creativity by requiring justification, while simultaneously enabling innovation by keeping alternative concepts in play. Research on design argumentation suggests that the quality of design outcomes is closely linked to the quality of argumentative exchanges within teams, particularly when ethical and stakeholder considerations are explicitly included (Lee et al., 2024). From a professional preparation perspective, learning to engage in such argumentative reasoning is central to becoming a competent engineer or

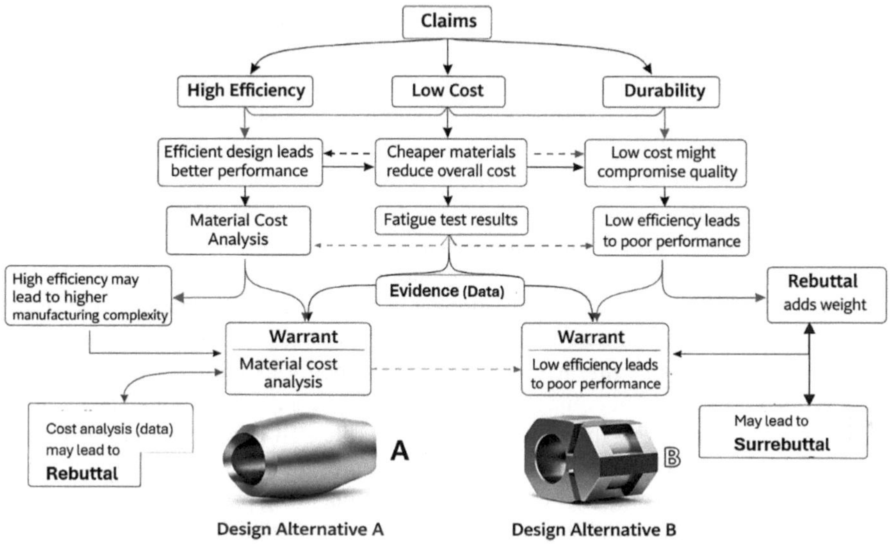

Fig. 7.2 Argumentation structure in engineering concept evaluation

technologist. It involves not only technical knowledge, but also rhetorical skill, epistemic humility, and ethical awareness.

Figure 7.2 invites us to look closely at how engineering concept evaluation is actually carried out during the concept development stage, making visible the Toulmin-inspired argumentative structure that often operates implicitly in professional practice. As we move through the figure, we see that concept development is not a matter of simply selecting the "best" idea from a list. Rather, we are drawn into a disciplined process of comparison in which competing design alternatives (A and B) are advanced, examined, and justified through evidence-based reasoning. At the top of the figure, we articulate claims around core design criteria, high efficiency, low cost, and durability. In doing so, we foreground the way engineers frame judgment in relation to explicitly defined requirements. These claims remind us that design teams do not treat concepts as neutral possibilities; instead, we actively argue for why one option may be preferable to another, given the constraints and priorities of the problem space. As we move downward, we identify the evidence that gives substance to these claims, including material cost analyses and fatigue test results. Here, the figure carries us into the practical heart of professional reasoning, where claims are supported, or challenged, by modelling outputs, empirical data, and feasibility analyses. The visual structure reinforces an important point: evidence does not speak for itself. We selectively mobilise data in relation to particular claims and design alternatives, shaping how that evidence informs judgment.

The figure then draws our attention to warrants, making explicit the engineering reasoning that links evidence to claims. Warrants such as material cost analysis or the principle that low efficiency leads to poor performance reflect the standards, principles, and disciplinary knowledge we rely on often tacitly during design reviews. By rendering these warrants visible, the figure aligns closely with our argument that professional judgment depends not only on data, but on shared understandings of what counts as valid reasoning in engineering. Crucially, Fig. 7.2 also foregrounds rebuttals and surrebuttals, reminding us that engineering claims are always provisional. By anticipating issues such as increased manufacturing complexity or added weight, we acknowledge limitations, risks, and trade-offs rather than concealing them. In this way, the figure reinforces our position that competent designers do not seek certainty where it cannot exist; instead, we reason transparently under conditions of uncertainty, attending to failure modes and unintended consequences. By placing rebuttals alongside both design alternatives, the figure carries us toward a collaborative view of argumentation. Competing concepts remain in play, their strengths and weaknesses openly examined. Argumentation here is not adversarial but constructive, functioning as a quality-control mechanism that disciplines creativity through justification while still leaving room for innovation. In this sense, Fig. 7.2 offers a concrete representation of how claims, data, warrants, and rebuttals interact in professional engineering practice, reinforcing our central argument that learning to engage in such structured argumentative reasoning is fundamental to becoming a competent, ethical, and reflective engineer.

7.4 Prototyping and Modelling Creativity in Engineering Design

Creativity in engineering is often romanticized as spontaneous insight. In professional practice, however, creativity is enacted through systematic representational work: modelling, simulation, prototyping, and visualization (Cross, 2008). These representations enable engineers to explore possibilities, test assumptions, and reveal emergent behaviours that cannot be deduced analytically. We distinguish among exploratory, evaluative, and communicative prototypes, each serving a different epistemic function within the design process. Exploratory prototypes expand the solution space by making abstract ideas tangible and provoking new questions. Evaluative prototypes test performance against explicit criteria and constraints. Communicative prototypes serve as boundary objects that facilitate dialogue among engineers, clients, users, and other stakeholders (Dym et al., 2005). Creativity emerges through iterative engagement with these representations. Unexpected failures, anomalies, or user responses are not merely obstacles; they are epistemic resources that drive innovation. From this perspective, creativity is less about novelty for its own sake and more about responsiveness to evidence, context, and feedback.

To understand how ethical reasoning operates within early-stage design work, it is therefore necessary to look closely at modelling and prototyping practices themselves. Figures 7.3and 7.4 show that ethical reasoning is embedded even at this stage of the design process. Choices about materials, data sources, testing environments, and user assumptions reflect judgments about safety, sustainability,

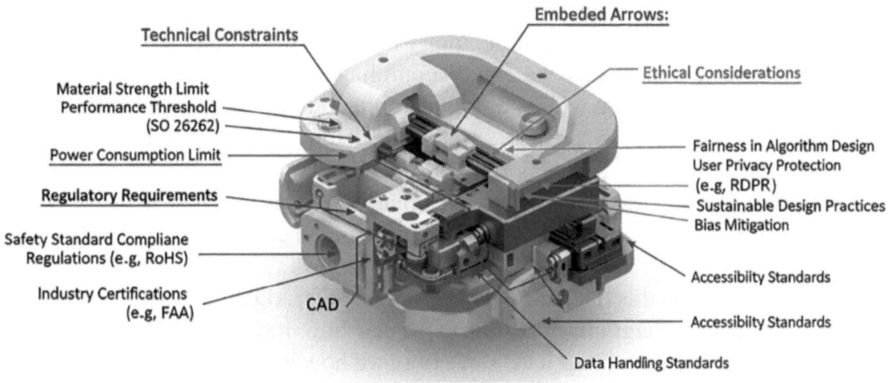

Fig. 7.3 Engineering Graphic 6.1 illustrating an annotated CAD model and system architecture

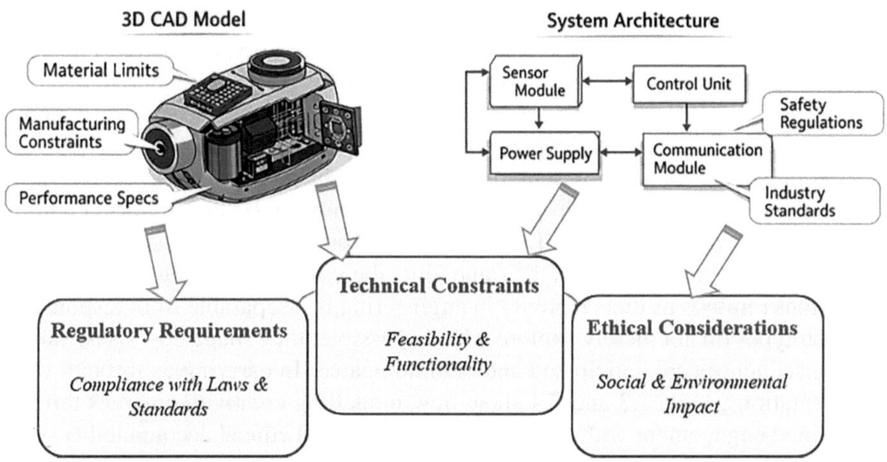

Fig. 7.4 Engineering Graphic 6.1- prototyping as a bridge between design intent and real-world constraints

inclusivity, and risk (Chivukula et al., 2021; Martin et al., 2021). For instance, the decision to prototype with synthetic data rather than real user data may mitigate privacy risks but limit ecological validity. Such trade-offs require explicit justification. Professional designers therefore treat prototypes not only as technical artifacts, but as ethical instruments that shape how risks and benefits are explored and communicated. Creativity, in this sense, is inseparable from responsibility as illustrated in Figs. 7.3 and 7.4. The diagrams show how creative exploration is disciplined through modelling and prototyping, where ethical reasoning and technical judgment are argued into the design rather than added afterward.

Figure 7.3 frames the CAD model as an argumentative space where technical constraints, regulatory requirements, and ethical considerations are integrated into modelling decisions and justified through annotation. As we trace the annotations, we see how material strength limits, power consumption thresholds, and performance standards shape component geometry and system layout. At the same time, regulatory requirements, such as safety compliance, industry certifications, and data-handling standards are woven into the architecture of the design itself. Ethical considerations are similarly present, guiding decisions about accessibility, privacy protection, sustainability, and bias mitigation. The diagram makes visible what is often implicit in professional practice: that creative exploration is structured by layered constraints, and that these constraints function as generative resources rather than mere limitations.

Figure 7.4 extends this insight by juxtaposing a 3D CAD model with a simplified system architecture diagram, allowing us to see how modelling and prototyping operate across representational levels. We observe how material limits, manufacturing constraints, and performance specifications inform physical design choices, while system-level considerations, such as sensor integration, power supply, control units, and communication modules are shaped by safety regulations and industry standards. The converging arrows toward technical constraints, regulatory requirements, and ethical considerations emphasize that exploratory, evaluative, and communicative prototypes are continuously informed by feasibility, compliance, and social impact. In this way, the figure illustrates creativity as an iterative dialogue between representations and judgments.

Together, these visual representations reinforce our argument that prototyping is an epistemic practice through which engineers think, test, and reflect. Failures, trade-offs, and tensions revealed through these representations become sources of insight that drive innovation. At the same time, the visual embedding of ethical considerations reminds us that creativity in engineering is inseparable from responsibility. Prototypes do not merely explore what is possible; they shape how risks, values, and consequences are anticipated and communicated. In carrying us through these representations, Figs. 7.3 and 7.4 show how modelling creativity emerges through disciplined engagement with evidence, constraints, and ethical accountability.

7.5 Iterative Design, Decision-Making, and Testing

Engineering design advances through iteration rather than linear progression. Each cycle of testing and refinement produces new evidence that may confirm, qualify, or overturn earlier decisions. Professional judgment lies in determining when further iteration enhances system integrity and when redesign is ethically or technically necessary (Simon, 1996). Decision-making at this stage integrates quantitative metrics, such as efficiency, reliability, and cost, with qualitative judgments concerning user experience, environmental sustainability, and social acceptability (NAE, 2020). These judgments are rarely reducible to optimization alone. Instead, they involve negotiated trade-offs among competing values and constraints. Argumentation again plays a central role. Design teams must justify why certain performance thresholds are deemed acceptable, why specific risks are tolerated, and why

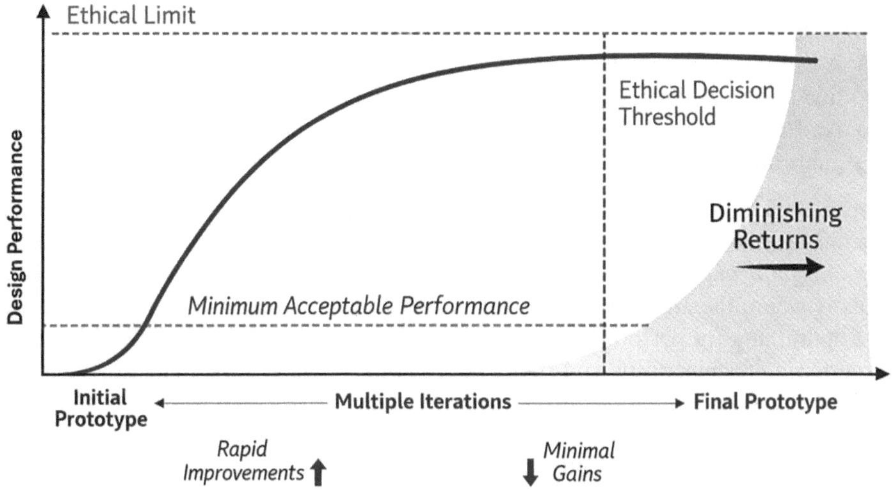

Fig. 7.5 Graph 6.1 design performance versus iteration cycles illustrating diminishing returns and ethical decision thresholds

alternative pathways are abandoned. Transparent documentation of these decisions supports accountability, enables knowledge transfer, and facilitates post-deployment evaluation, particularly in high-stakes systems where failures carry significant societal consequences (Van de Poel et al., 2018).

Explicit documentation of design rationales is essential. Such records support accountability, enable knowledge transfer, and facilitate post-deployment evaluation, particularly in high-stakes systems where failures carry significant societal consequences. Iteration also has an ethical dimension. Continuing to iterate on a flawed design may expose users or communities to unnecessary risk, while premature closure may foreclose safer or more equitable alternatives. Professional responsibility involves recognizing these tensions and making them explicit within decision-making processes. In Fig. 7.5 we see a visual representation of how iteration shapes design performance over time, and how ethical judgment must guide when we stop refining a prototype. The graph plots *Design Performance* on the vertical axis against *Iteration Cycles* on the horizontal axis. We begin at the Initial Prototype, where performance is low and improvements are rapid. The curve rises steeply in this early phase, showing that early iterations produce significant gains with relatively little effort. This matches our earlier discussion that iteration is not linear but evidence driven. Each test generates new information that quickly improves the design. As we move into Multiple Iterations, the curve begins to flatten. Performance continues to improve, but the gains become smaller with each cycle. This flattening signals the onset of diminishing returns, which is highlighted by the shaded region on the right side of the graph. Here we see why professional judgment matters. Even though performance is still improving, the benefits per iteration are shrinking, and further refinement may no longer be justified.

The graph also introduces two crucial decision thresholds:

(i) A Minimum Acceptable Performance line (dashed brown), indicating the baseline level that the design must meet before it can be considered viable.
(ii) An Ethical Limit line (dashed green), showing an upper bound where further gains may require disproportionate cost, risk, or ethical compromise.

Between these two lines is a red dashed "Ethical Decision Threshold", a point where we must pause and ask: *Is continued iteration adding value, or are we simply chasing marginal improvements while exposing users or society to unnecessary risk?* This is where the argument about negotiated trade-offs becomes concrete. We are not optimizing for performance alone; we must also consider cost, reliability, user experience, sustainability, and social acceptability. The curve's flattening near the ethical threshold reminds us that iteration is not merely a technical process. It is a moral one. We must decide when further testing is justified and when it becomes ethically questionable to continue. In this way, the figure supports our claim that professional responsibility involves recognizing the tension between refinement and risk, and making those decisions explicit and defensible through argumentation and documentation. In short, Fig. 7.5 helps us conceptualize the iterative design process, illustrating how early gains give way to diminishing returns and necessitate ethical judgment before a final prototype is reached, thereby reinforcing iteration as an evidence-informed and responsible practice.

7.6 Ethics, Communication, and the Defence of Design Models in Engineering

Ethical reasoning in engineering design is not an afterthought applied once technical decisions are complete. It is interwoven throughout the design process. Engineers must anticipate how their designs shape human behaviour, redistribute risks and benefits, and transform social and ecological systems (Martin et al., 2021; Winner, 1986). Ethical responsibility is enacted through communication. Practitioners are accountable not only for what they design, but for how they explain, justify, and defend their models to regulators, clients, users, and the broader public. Transparent argumentation fosters trust, enables informed decision-making, and supports democratic engagement with technology (Erduran et al., 2015). Socio-STEM issues, such as algorithmic bias, environmental justice, data privacy, and sustainability, demand that engineers articulate the ethical grounds of their decisions explicitly. Frameworks such as value-sensitive design and ethics-by-design provide structured approaches for integrating values into technical decision-making. However, these frameworks are effective only insofar as they are supported by genuine communicative practices that invite scrutiny and dialogue. Professional integrity therefore requires acknowledging uncertainty, disclosing limitations, and resisting pressures to obscure risk or

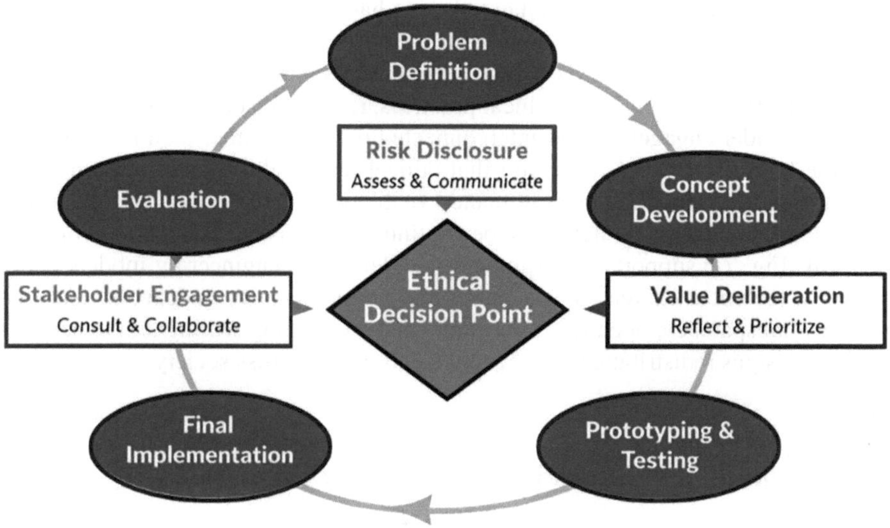

Fig. 7.6 Ethical decision points across the engineering design cycle

oversell technological solutions (Spiekermann & Winkler, 2020). The defence of design models is ultimately a moral as well as a technical endeavour.

Figure 7.6 illustrates the heart of the argument developed in this section by making explicit that ethical reasoning and communication are not peripheral to engineering design but are structurally embedded within it. As we move with the diagram around the cyclical design process, from problem definition through concept development, prototyping, implementation, and evaluation, we see that ethical responsibility does not reside at a single stage. Instead, the figure positions ethical decision points at the centre of the cycle, reminding us that each phase invites scrutiny, justification, and dialogue. In other words, the central ethical decision point signals that design choices are always provisional and contestable, requiring justification as new evidence, constraints, and consequences emerge. By foregrounding stakeholder engagement, risk disclosure, and value deliberation at key moments in the cycle, the figure illustrates how ethical responsibility is enacted through communication. As we move further through the cycle, stakeholder engagement becomes particularly prominent during evaluation, underscoring that ethical responsibility is enacted through consultation and collaboration rather than unilateral expertise. Here, we are reminded that engaging stakeholders involves more than consultation. It is an ongoing dialogue that shapes how problems are framed and how solutions are evaluated.

Similarly, risk disclosure is presented not as a compliance exercise but as a communicative act through which uncertainty, limitations, and potential harms are openly assessed and conveyed. In this sense, risk disclosure emerges early, at the point of problem definition, where engineers must already begin to assess and communicate uncertainties rather than defer them. As concepts are developed, our attention is drawn to value deliberation, signalling that technical optimisation is inseparable from reflective judgement about what should be prioritised, protected, or traded off. Value deliberation highlights the need to reflect on which interests are prioritised and why, making visible the normative assumptions embedded in technical models. By arranging these ethical practices, risk disclosure, value deliberation, and stakeholder engagement, as interconnected moments rather than isolated add-ons, the figure reinforces our claim that ethical responsibility is enacted through communication. The circular flow makes visible how explanations, justifications, and defences of design models must be continuously revisited as designs evolve. In this way, Fig. 7.6 supports the argument that defending engineering models is not merely a technical exercise but a moral and communicative endeavour, one that requires transparency, openness to critique, and an ongoing willingness to account for how designs redistribute risks, benefits, and values across society.

7.7 Socio-STEM Issues in Engineering Design Tasks

In the foregoing sections, we have argued that engineering design thinking is fundamentally creative, ethical, and socially situated. We now extend this argument by foregrounding Socio-STEM issues as an integral dimension of engineering design practice. We proceed from the premise that engineering problems are never purely technical; rather, they are socio-technical configurations in which material artefacts, human values, institutional structures, and ecological systems are entangled (Bijker et al., 2012; Latour, 2005). As such, every engineering task simultaneously constitutes a technical intervention and a social proposition.

When we engage students and practitioners in engineering design tasks, we are therefore also inviting them to engage with questions of power, responsibility, risk, and justice (Bucciarelli, 2008; NAE, 2014). Issues such as algorithmic bias, energy transitions, water security, data privacy, infrastructure inequality, biomedical innovation, and environmental sustainability are not external "contexts" applied after design decisions are made; they are constitutive of what it means to design responsibly in contemporary STEM fields (NAE, 2020; Stilgoe et al., 2013). Rather than framing tasks solely in terms of optimisation or efficiency, we invite students, practitioners, and emerging professionals to ask: Who benefits? Who bears the risk? What values are embedded in this solution? What alternative futures does this design enable or foreclose?

In practical terms, this means that engineering tasks should explicitly require:

(i) Problem framing informed by social context, where societal needs, historical inequities, and environmental constraints are articulated alongside technical specifications.
(ii) Evidence-based reasoning, drawing on scientific data, engineering models, and socio-ethical analysis.
(iii) Value deliberation, where competing priorities (e.g., cost, safety, sustainability, accessibility) are made explicit and justified.

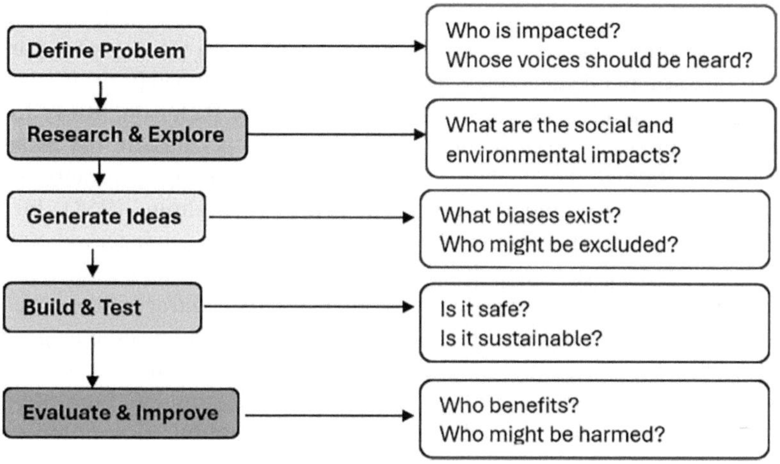

Fig. 7.7 Socio STEM infused engineering design task

(iv) Stakeholder engagement, recognising users, communities, regulators, and non-human stakeholders as legitimate participants in the design process.

Rather than presenting engineering design as a linear technical routine, Fig. 7.7 draws us into a guided inquiry in which each familiar design stage is deliberately aligned with socio-ethical questions that surface the social meaning of design decisions from the outset (Dym et al., 2005). As we move from defining the problem to evaluating and improving solutions, we see that each technical phase is paired with a reflective prompt that invites us to pause and interrogate the social consequences of our choices. At the outset, problem definition is no longer treated as a neutral technical act; instead, we are asked to consider who is impacted and whose voices ought to be heard, underscoring that framing itself is already an ethical and political decision.

As we proceed to research and exploration, attention shifts beyond technical feasibility to encompass social and environmental impacts, reinforcing the claim that evidence in engineering design must be both scientific and socio-ethical (Sadler, 2011). Idea generation is similarly reframed as a site where assumptions, biases, and exclusions may arise, prompting us to reflect on who might be marginalised by particular design pathways (Bucciarelli, 2008). During the build and test phase, questions of safety and sustainability foreground responsibility and risk, reminding us that performance cannot be separated from care for people and environments (NAE, 2014). Finally, evaluation and improvement compel us to assess not only whether a solution works, but also how its benefits and harms are distributed (Stilgoe et al., 2013). In this way, engineering design tasks emerge as inherently socio-technical configurations rather than purely technical exercises, carrying learners into a sustained inquiry where technical reasoning, ethical deliberation, and social accountability unfold together across the design process (Bijker et al., 2012). Once design is understood in these terms, the judgments it entails cannot remain implicit or procedurally assumed; they must be articulated, defended, and open to critical scrutiny (Osborne, 2010).

Socio-STEM issues therefore demand that engineering reasoning be expressed through structured argumentation, enabling designers to justify why particular trade-offs are acceptable and others are not, and to move beyond procedural compliance toward communicative accountability (Osborne & Patterson, 2011; Toulmin, 1958). We emphasise argumentation not as rhetorical persuasion, but as an epistemic practice through which claims are supported by evidence, warrants are made explicit, and counter-arguments are taken seriously (Toulmin, 1958). In Socio-STEM contexts, ethical reasoning becomes visible when engineers explain how uncertainty is managed, how risks are disclosed, and how moral responsibility is distributed across systems and institutions (NAE, 2020). Contrary to the misconception that ethics constrains creativity, Socio-STEM issues often serve as generative constraints that expand the design space (Stilgoe et al., 2013). Sustainability requirements, equity considerations, and safety standards can stimulate innovative solutions that would not emerge under purely technical optimisation criteria (NAE, 2014). In light of this, practitioners and future engineers are encouraged to view ethical and social constraints as productive design parameters. Responsible innovation emerges when creativity is exercised within an awareness of long-term societal consequences and planetary boundaries.

For students preparing to enter STEM professions, engagement with Socio-STEM issues is therefore not an optional enrichment activity, but a core professional competency (NAE, 2020; Sadler, 2011). Engineers are increasingly called upon to justify their designs to diverse audiences, operate under public scrutiny, and contribute to democratic deliberation about technology (Winner, 1986). By embedding Socio-STEM issues into engineering design tasks, we cultivate practitioners who are not only technically proficient but also ethically literate, communicatively

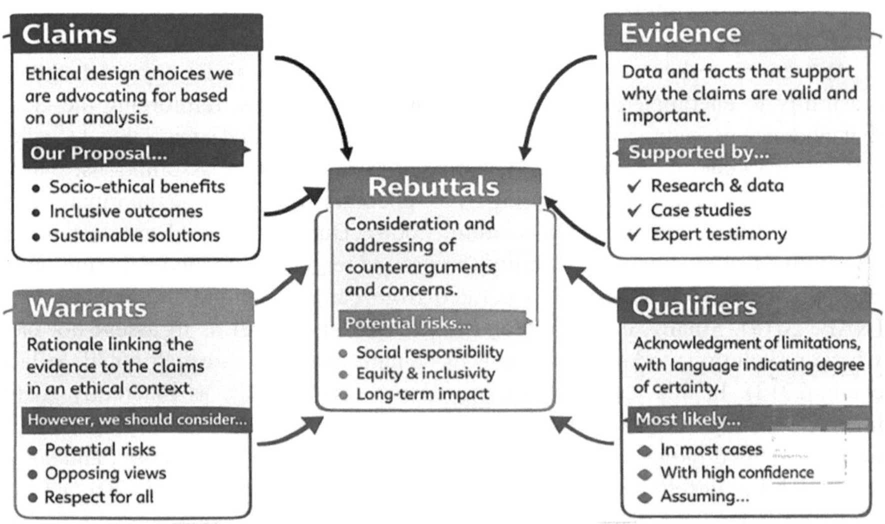

Fig. 7.8 Toulmin inspired argumentation structure for socio STEM design decisions

competent, and socially responsive (Osborne, 2010). Figure 7.8 extends our inquiry by making visible the internal structure of the reasoning through which Socio-STEM design decisions are ethically justified. Drawing on a Toulmin-inspired argumentation framework, the design idea is framed as a claim, an explicit proposal about what ought to be done and why, signalling that engineering decisions are always assertions that invite justification (Toulmin, 1958). Evidence, warrants, qualifiers, and rebuttals collectively foreground ethical reasoning, uncertainty, and dialogic engagement, reinforcing the view that responsible engineering design is revisable, transparent, and accountable rather than final or absolute (Osborne & Patterson, 2011).

As we follow the logic further, the warrants linking evidence to claims foreground the ethical reasoning that is often left implicit in engineering practice. Here, we see how values, assumptions, and normative commitments function as bridges between what is known and what is proposed. At the same time, the inclusion of qualifiers prompts us to acknowledge uncertainty and degrees of confidence, reinforcing the idea that responsible engineering involves transparency about limitations rather than overstated certainty. The explicit positioning of rebuttals invites us to anticipate and engage counterarguments, risks, and competing perspectives, emphasising that ethical reasoning in Socio STEM contexts is dialogic and revisable rather than final or absolute. Taken together, argumentation emerges here as an epistemic–ethical practice through which Socio STEM design decisions are rendered intelligible, defensible, and open to revision. In this light, communicative accountability stands alongside technical proficiency as a core dimension of engineering expertise, while creativity is refined through the justification of trade-offs and sustained attention to equity, sustainability, and long-term consequences. In this way, the representation supports our broader argument that embedding structured argumentation into design tasks prepares students and practitioners to defend their decisions, manage uncertainty, and act responsibly within complex social and institutional environments.

7.8 Chapter Summary

In this chapter, we positioned engineering design thinking as a professional framework that integrates creativity, argumentation, and ethical reasoning within the practice of STEM. We demonstrated that engineering design is not merely a sequence of technical steps, but a disciplined process of judgment in which problem scoping, concept development, prototyping, iteration, and communication are guided by evidence, shaped by values, and evaluated in light of their broader societal and ecological implications. Throughout the design process, we saw how argumentation makes reasoning visible and accountable, enabling practitioners to justify decisions, evaluate alternatives, and refine solutions in response to both empirical findings and ethical considerations.

We further emphasised that developing competence in engineering design thinking requires more than technical proficiency. It calls for cultivating the intellectual dispositions necessary to engage uncertainty, anticipate consequences, and deliberate responsibly within complex socio-technical systems. As emerging professionals learn to navigate design challenges, they also learn to balance innovation with responsibility, ensuring that their solutions serve not only functional purposes but also human and environmental well-being. In this sense, engineering design becomes a form of ethical and epistemic practice, grounded in evidence, informed by dialogue, and oriented toward responsible action.

At the same time, we recognised that contemporary engineering practice increasingly unfolds within interconnected and rapidly evolving knowledge environments shaped by digital technologies, collaborative networks, and continuous information exchange. As practitioners engage with these networked contexts, their capacity to design responsibly depends not only on how they generate solutions, but also on how they learn, evaluate knowledge, and participate in distributed communities of practice. This recognition invites us to consider the broader learning frameworks that support critical, creative, and ethical engagement in such environments, which we take up in the next chapter through an examination of connectivist learning as a foundation for responsible and adaptive STEM professionalism.

7.8.1 Looking Ahead to Chap. 8

Having established engineering design thinking in Chap. 7 as a professional practice grounded in creativity, argumentation, and ethical responsibility, we now recognise that such practice does not occur in isolation. We have seen how design decisions emerge through processes of evidence-based reasoning, iterative refinement, and ethical foresight, requiring practitioners to justify their choices and remain accountable to both society and the natural world. This positioning challenges us to understand engineering not simply as a technical endeavour, but as a form of professional judgment exercised within broader socio-technical and knowledge systems. As STEM environments become increasingly shaped by digital connectivity, intelligent technologies, and rapidly evolving knowledge, we must also consider how practitioners learn, adapt, and reason within these dynamic and networked contexts.

Chapter 8 takes up this challenge by introducing the connectivist learning framework as a guide for cultivating critical, creative, and ethical STEM professionals in complex, interconnected environments. We examine how learning unfolds across networks of people, technologies, and information systems, and how practitioners develop the capacity to navigate these distributed knowledge spaces responsibly. We explore key connectivist principles, including diversity, autonomy, openness, adaptability, and continual knowledge renewal, and consider how these principles support sustained professional growth in fields characterised by uncertainty and rapid change.

Importantly, we also examine how argumentation and ethical reasoning remain essential within networked learning environments. We demonstrate how argumentation provides a means of evaluating knowledge claims, resolving tensions among competing perspectives, and maintaining epistemic integrity in contexts where information is abundant but uneven in quality. Through this lens, we extend our understanding of professional responsibility beyond the design of technological solutions to include participation in the ongoing construction and regulation of knowledge itself. In this way, Chap. 8 broadens the scope of our inquiry, equipping us with a framework for learning, reasoning, and acting as ethically grounded and intellectually agile professionals within an increasingly connected STEM landscape.

7.9 Reflective Prompts for Professional Practice

1. How do our problem framings implicitly privilege certain values, stakeholders, or outcomes?
2. In what ways does structured argumentation enhance rather than constrain creative design work?
3. How can design documentation function as an ethical artifact across the lifecycle of a technology?
4. How should engineers respond when technical optimization conflicts with societal or environmental values?

References

Baldassarre, B., Calabretta, G., Karpen, I. O., Bocken, N., & Hultink, E. J. (2024). Responsible design thinking for sustainable development: A critical literature review and research agenda. *Journal of Business Ethics, 195*, 25–46. https://doi.org/10.1007/s10551-023-05600-z

Bijker, W. E., Hughes, T. P., & Pinch, T. J. (Eds.). (2012). *The social construction of technological systems: New directions in the sociology and history of technology* (Anniversary ed.). MIT Press.

Bucciarelli, L. L. (2008). Ethics and engineering education. *European Journal of Engineering Education, 33*(2), 141–149. https://doi.org/10.1080/03043790801979856

Chivukula, S. S., Li, Z., Pivonka, A. C., Chen, J., & Gray, C. M. (2021). Surveying the landscape of ethics-focused design methods. *ACM Computing Surveys, 54*(7), 1–36. https://doi.org/10.1145/3468264

Cross, N. (2008). *Engineering design methods: Strategies for product design* (4th ed.). Wiley.

Dym, C. L., Agogino, A. M., Eris, O., Frey, D. D., & Leifer, L. J. (2005). Engineering design thinking, teaching, and learning. *Journal of Engineering Education, 94*(1), 103–120. https://doi.org/10.1002/j.2168-9830.2005.tb00832.x

Erduran, S., Osborne, J., & Simon, S. (2015). The role of argumentation in developing scientific literacy. *Science Education, 99*(5), 877–902. https://doi.org/10.1002/sce.21170

Iwuanyanwu, P. N. (2020). Nature of problem-solving skills for 21st Century STEM Learners: What teachers need to know. *Journal of STEM Teacher Education, 55*(1), 27 – 40. https://doi.org/10.30707/JSTE55.1/MMDZ8325

Latour, B. (2005). *Reassembling the social: An introduction to actor-network-theory*. Oxford University Press.

Lee, W., Mentzer, N., Jackson, A., Bartholomew, S., & Clevenger, A. (2024). Defining and evaluating argumentation quality in the context of design thinking. *Design and Technology Education: An International Journal, 29*(3), 11–35.

Martin, D. A., Conlon, E., & Bowe, B. (2021). A multi-level review of engineering ethics education: Towards a socio-technical orientation. *Science and Engineering Ethics, 27*(5), 60. https://doi.org/10.1007/s11948-021-00333-6

National Academy of Engineering. (2020). *Ethics, equity, and the future of engineering*. National Academies Press.

National Academy of Engineering & National Research Council. (2014). *STEM integration in K–12 education: Status, prospects, and an agenda for research*. National Academies Press. https://doi.org/10.17226/18612

Ogunniyi, M. B. (2007). Teachers' stances and practical arguments regarding a science–indigenous knowledge curriculum: Part 1. *International Journal of Science Education, 29*(8), 963–986. https://doi.org/10.1080/09500690600931020

Osborne, J. (2010). Arguing to learn in science: The role of collaborative, critical discourse. *Science, 328*(5977), 463–466. https://doi.org/10.1126/science.1183944

Osborne, J., & Patterson, A. (2011). Scientific argument and explanation: A necessary distinction? *Science Education, 95*(4), 627–638. https://doi.org/10.1002/sce.20438

Sadler, T. D. (Ed.). (2011). *Socio-scientific issues in the classroom: Teaching, learning and research*. Springer. https://doi.org/10.1007/978-94-007-1159-4

Simon, H. A. (1996). *The sciences of the artificial* (3rd ed.). MIT Press.

Spiekermann, S., & Winkler, T. (2020). Value-based engineering for ethics by design. *Information Systems Frontiers, 22*(5), 1037–1051. https://doi.org/10.1007/s10796-019-09990-7

Stilgoe, J., Owen, R., & Macnaghten, P. (2013). Developing a framework for responsible innovation. *Research Policy, 42*(9), 1568–1580. https://doi.org/10.1016/j.respol.2013.05.008

Toulmin, S. E. (1958). *The uses of argument*. Cambridge University Press.

Van de Poel, I., Mehos, D. C., & Asveld, L. (Eds.). (2018). *New perspectives on technology in society: Experimentation beyond the laboratory*. Routledge.

Winner, L. (1986). *The whale and the reactor: A search for limits in an age of high technology*. University of Chicago Press.

Zeidler, D. L. (2014). Socioscientific issues as a curriculum emphasis: Theory, research and practice. In N. G. Lederman & S. K. Abell (Eds.), *Handbook of research on science education* (Vol. 2, pp. 697–726). Routledge.

Chapter 8
Connectivist Learning Framework for Critical, Creative, and Ethical STEM Minds

8.1 Introduction

In the preceding chapters, we have argued that argumentation is not an ancillary skill in STEM, but the epistemic engine that drives scientific inquiry, engineering design, and responsible technological decision-making. In this final chapter, we draw these strands together by examining connectivism as a learning framework suited to contemporary STEM conditions, conditions marked by distributed expertise, digital mediation, and rapidly evolving knowledge claims. In today's STEM environments, what matters is no longer simply what one knows, but how we connect; how we locate and evaluate information, interrogate competing claims, collaborate across disciplinary boundaries, and revise judgments as systems and evidence evolve. Connectivism offers a powerful lens for understanding this shift by framing learning as participation in networks of people, tools, data, and ideas (Bell, 2011; Siemens, 2005). From this perspective, knowing is not confined to the individual mind but distributed across socio-technical systems that extend cognition beyond the learner.

Yet this very strength has also generated sustained critique. Critics have argued that connectivism risks collapsing epistemic standards, mistaking access for understanding, and offering description without sufficient normative guidance for judgment and responsibility (Clarà & Barberà, 2013; Goldie, 2016). We take these critiques seriously. Rather than defending connectivism as a complete learning theory, this chapter advances it as a professional epistemic framework, one whose limitations can be addressed only when its processes are disciplined through argumentation. Argumentation, we contend, provides the normative structure that connectivism itself does not specify. It is through practices of justification, critique, and reasoned decision-making that networked learning becomes epistemically defensible, creatively generative, and ethically grounded. By foregrounding argumentation, we situate

P. N. Iwuanyanwu, *Empowering STEM Thinkers Through Argumentation*, Contemporary Trends and Issues in Science Education 69,
https://doi.org/10.1007/978-3-032-27906-4_8

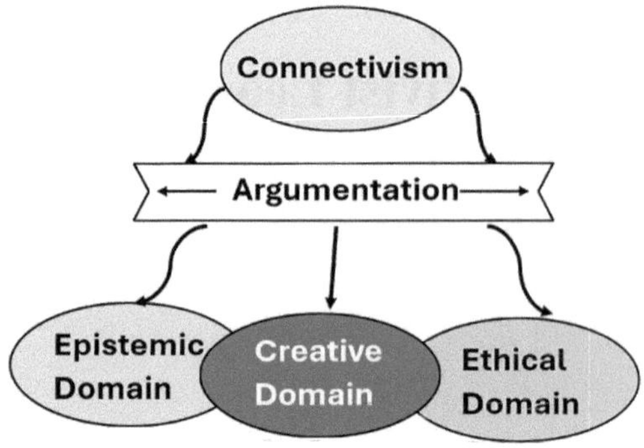

Fig. 8.1 A conceptual framework positioning argumentation as the regulating structure of connectivism

connectivism within a broader philosophical account of STEM reasoning that emphasises judgment, accountability, and ethical agency under conditions of uncertainty.

Our purpose in this chapter is threefold. First, we articulate the core principles of connectivism in ways that speak directly to STEM practice rather than abstract learning theory. Second, we show how argumentation operates within networked environments to regulate professional reasoning, innovation, and ethical action. Third, we offer design-oriented guidance for cultivating connectivist STEM learning experiences that prepare students not merely to enter STEM professions, but to participate in shaping them responsibly. Figure 8.1 frames the central inquiry of this chapter: *in what sense can connectivism move beyond description to function as a disciplined framework shaped through argumentation?*

At the top of the diagram, Connectivism sits as the central organising idea, reflecting our earlier claim that in contemporary STEM contexts, learning is less about what an individual knows and more about how they connect, evaluate, and update knowledge within distributed networks. From this central concept, arrows flow downward into a banner labelled Argumentation, signalling that connectivist learning does not proceed unstructured. Rather, it is mediated by the practices of reasoning, justification, and critique that argumentation provides. This positioning mirrors our chapter's purpose that aim to take the criticisms of connectivism seriously and show that its limitations, such as risks of relativism or uncritical information access, are addressed when we treat argumentation as the normative mechanism that governs networked learning.

Below the argumentation layer, three overlapping domains emerge: (1) Epistemic, (2) Creative, and (3) Ethical. These domains are intentionally placed together to show that connectivism, when regulated by argumentation, extends across multiple dimensions of STEM practice.

(i) Epistemic domain represents the work of evaluating claims, verifying evidence, and distinguishing credible knowledge from unreliable information, exactly the epistemic discipline we argue is essential in digitally mediated, networked systems.

(ii) Creative domain captures the innovation and design-oriented reasoning that STEM professionals must engage in, but which must be grounded in justifiable reasoning rather than speculative novelty.
(iii) Ethical domain highlights the responsibility and moral judgment required when technology affects society, aligning with our insistence that connectivism alone cannot account for the ethical dimension of STEM decision-making.

By situating Argumentation as the bridge between *Connectivism* and these three domains, we explore the conditions under which connectivist learning can be rendered intellectually defensible and ethically grounded. The diagram thus supports a movement from critique to synthesis, showing that networked knowledge, innovation, and responsibility are integrated into a coherent framework for STEM learning and professional practice only when connectivist processes are disciplined through argumentation.

8.2 Key Principles of Connectivism

Connectivism conceptualises learning as the capacity to establish and traverse networks of information, expertise, and practice. Its core principles, diversity, autonomy, openness, adaptability, and continual knowledge updating, align closely with the epistemic demands of contemporary STEM work (Siemens, 2005). In this regard, several interrelated principles of connectivism are particularly salient for STEM professionals. As argued in Chap. 6, socio-STEM issues demand engagement with heterogeneous forms of evidence. For instance, continual knowledge updating emerges from exposure to multiple perspectives, data sources, disciplinary lenses, and cultural contexts. However, each principle of connectivism also introduces epistemic risks that must be explicitly managed. Diversity expands the range of perspectives and data available for reasoning, supporting creativity and robust problem framing. At the same time, unregulated diversity can produce fragmentation, misinformation, and false equivalence among claims. Argumentation mitigates this risk by requiring explicit justification, evidentiary standards, and critical comparison of competing positions (Osborne et al., 2004; Toulmin, 1958). It is on this view that connectivism treats diversity not as noise to be filtered out, but as a condition for robust reasoning.

Within STEM environments, connectivism can provide an enabling intellectual space for learner and practitioner autonomy in selecting networks and sources. However, autonomy must therefore be coupled with accountability, professionals are responsible not only for their conclusions, but for the justificatory pathways through which those conclusions are reached (Kuhn, 1993). Open systems accelerate innovation and collective problem solving, yet they also amplify exposure to unreliable information. Argumentation functions here as an epistemic filter, enabling practitioners to interrogate credibility, assess methodological rigor, and resist the normalisation of weak claims. The capacity to revise beliefs in response to new evidence is essential in fast-moving STEM fields. However, continual updating without evaluative criteria risks epistemic instability. Structured argumentation stabilises adaptation by grounding revision in evidence rather than novelty. Connectivism's claim that knowledge resides in networks rather than individuals has been criticised for dissolving responsibility (Garcia et al., 2015; Shrivastava, 2018). This chapter argues instead that distributed knowing intensifies ethical responsibility, as professionals must justify how networked inputs are selected, weighted, and acted upon.

In rapidly evolving STEM fields, knowledge has a short half-life. Learning, therefore, involves continual updating of beliefs in response to new evidence, an extension of the iterative reasoning cycles discussed in Chap. 5. No single individual holds complete expertise. Professional competence resides in the network itself, requiring collaborative argumentation and negotiated meaning-making. Table 8.1 extends this analysis by making explicit the relationship between core connectivist principles, the epistemic risks they introduce in networked learning environments, and the regulatory role of argumentation. Rather than treating connectivism as an unproblematic account of learning in digital and professional networks, we use the table to interrogate what is gained, and what is put at risk, when learning is organised around distributed, dynamic, and open systems.

Table 8.1 Connectivist principles, associated epistemic risks, and the regulatory role of argumentation

Connectivist principle	Epistemic risk in networked learning	Regulatory role of argumentation
Knowledge is distributed across networks of people, tools, and information	Epistemic relativism, where all sources are treated as equally valid regardless of credibility	Argumentation establishes standards for evidence, warrants, and source credibility, enabling reasoned judgment about competing knowledge claims
Learning consists in forming and navigating connections	Fragmentation and incoherence arising from exposure to large volumes of unintegrated information	Argumentation supports synthesis by requiring learners to connect claims logically and justify relationships among ideas
Capacity to recognise patterns across networks is more important than content mastery	Superficial pattern recognition and confirmation bias reinforced by algorithmic filtering	Argumentation introduces counter-claims, alternative explanations, and critical scrutiny of inferred patterns
Knowledge is dynamic, provisional, and continually evolving	Instability of understanding and premature acceptance of emerging or untested claims	Argumentation frames knowledge as revisable through critique, evidence evaluation, and reason-giving practices
Access to information is central to learning	False authority and overreliance on visibility, popularity, or algorithmic ranking	Argumentation shifts emphasis from access to justification, requiring claims to be defended with reasons and evidence

Across Table 8.1, we trace a recurring pattern showing that principles which enable learning in complex STEM contexts simultaneously introduce vulnerabilities that exceed the explanatory power of connectivity alone. Distributed knowledge, for example, expands epistemic reach but risks relativism unless standards of evidence and credibility are articulated. Similarly, the emphasis on forming and navigating connections supports adaptability, yet invites fragmentation unless relationships among ideas are justified and integrated. In each case, argumentation functions not as an external add-on but as a regulatory practice that stabilises meaning, supports judgment, and renders reasoning accountable. By juxtaposing principles, risks, and regulatory practices, Table 8.1 carries our inquiry from description to discipline. It shows how argumentation transforms diversity from epistemic noise into a resource for reasoning, autonomy into accountable agency, openness into critical scrutiny, and continual updating into evidence-based revision. In this way, the table reinforces our central claim that disciplined argumentative practices are essential for connectivism to become intellectually coherent and ethically defensible within STEM learning.

8.3 Learning Networks and Knowledge Flow

In connectivist environments, knowledge flows through networks rather than hierarchies. Information moves across institutional, disciplinary, and geographical boundaries, reshaping how STEM problems are framed and solved. From a professional standpoint, this shift reflects contemporary practice realities in which solutions emerge through distributed collaboration rather than isolated expertise. Learning networks may be conceptualised as configurations of human nodes (experts, peers, stakeholders), non-human nodes (databases, sensors, algorithms, instruments), and socio-technical structures (platforms, standards, regulatory systems). Within such configurations, argumentation functions as the epistemic mechanism that enables participants to assess the credibility of claims, justify decisions, and coordinate action across heterogeneous nodes.

A central argument in favour of connectivist knowledge flow is that it mirrors how contemporary STEM knowledge is produced: iteratively, collaboratively, and across institutional boundaries (Bell, 2011). Networked knowledge flow supports rapid problem redefinition and adaptive response, particularly in complex systems where no single actor holds sufficient information. However, critics argue that networked flow risks privileging connectivity over coherence, leading to shallow engagement, epistemic overload, and the amplification of unvetted claims (Goldie, 2016). This critique is well founded if knowledge flow is treated as self-regulating. The connectivist framework advanced here rejects that assumption. Argumentation provides the evaluative structure through which knowledge flow becomes disciplined rather than diffuse. By requiring explicit warrants, evidentiary standards, and critical comparison, argumentation enables practitioners to stabilise meaning within dynamic networks and to resist the normalisation of speed and consensus as

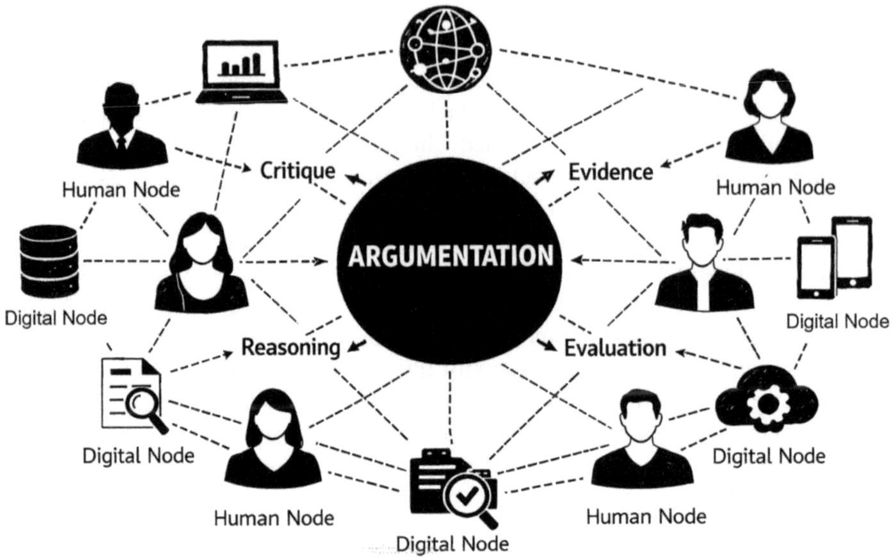

Fig. 8.2 Argumentation as the epistemic core of learning networks

substitutes for rigor. From a professional standpoint, the critical question is not whether information is available, but how it is filtered, validated, and integrated into decision-making processes. This echoes our discussion in Chap. 3 on evidence evaluation, now extended to large-scale, digitally mediated contexts.

Taken together, Fig. 8.2 progressively elaborate how knowledge flow in connectivist learning networks becomes intellectually and professionally viable only when regulated through argumentation. The figures synthesise these ideas into a balanced human–digital ecology in which connectivity is disciplined by epistemic practices rather than left to operate unchecked. Figure 8.2 establishes the foundational logic of disciplined knowledge flow in connectivist environments. Human and digital nodes are distributed around a central argumentative core, signalling that connectivity alone does not generate knowledge. Instead, argumentation is positioned as the organising epistemic force through which critique, evidence, reasoning, and evaluation circulate across the network. Its deliberate placement at the centre emphasises that connectivity, by itself, does not confer epistemic legitimacy. Human and digital nodes are arranged symmetrically around this core, underscoring that neither human expertise nor technological infrastructure is epistemically privileged in isolation. What matters is how claims, data, interpretations, and decisions move through argumentative practices such as critique, evidence evaluation, and reason-giving. The directional flows in the diagram further illustrate that knowledge does not travel linearly from source to user. Rather, it is continuously shaped through interaction among nodes that contribute different forms of expertise, data, and perspective.

Within this configuration, argumentation functions as an integrative mechanism that coordinates distributed inputs into defensible judgements. The figure thus

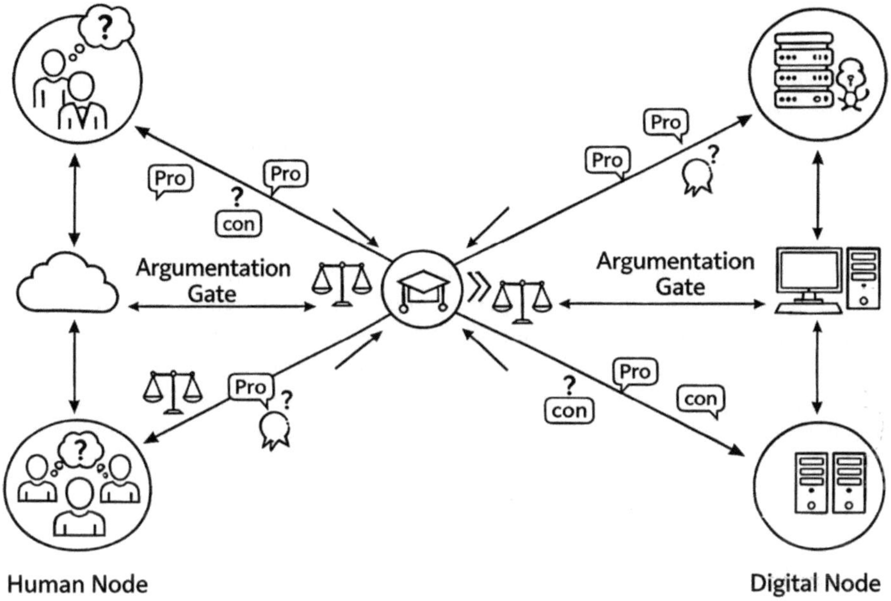

Fig. 8.3 Argumentation gates and the regulation of knowledge flow

makes explicit our claim that learning in contemporary STEM contexts is less about individual cognition and more about participation in disciplined epistemic networks. Multiple human nodes interact dynamically with digital repositories and systems, while argumentation explicitly regulates knowledge flow at critical points of exchange. In this way, connectivist learning becomes epistemically robust when argumentative practices mediate interaction, preventing diffusion, overload, or uncritical consensus. Knowledge is neither confined to individuals nor dissolved into networks; it is stabilised through sustained justification, critique, and coordination across the system.

Figure 8.3 advances this analysis by foregrounding the regulatory function of argumentation within the network. Rather than depicting free-flowing connectivity, the figure introduces argumentation gates that mediate exchanges between human and digital nodes. These gates symbolise moments at which claims must be justified, challenged, or revised before they are permitted to circulate further within the system. Here, argumentation gates operate explicitly between human and digital nodes, emphasising that information does not move freely or unexamined. Claims are weighed through pro–con structures, evidentiary checks, and critical judgement before being integrated into shared understanding or operational decision-making. This depiction directly addresses common critiques of connectivism by demonstrating that coherence is actively produced, not passively inherited, within networked learning environments. Knowledge flow is thus shown to be selective, structured, and accountable.

By visualising pro–con reasoning, evidentiary weighing, and critical judgement, the figure responds to concerns that connectivist learning risks epistemic relativism or informational overload. Knowledge circulation is conditional rather than automatic, with only those claims that withstand argumentative scrutiny becoming stabilised for further action. This representation reinforces our position that coherence in networked learning is constructed through evaluative practices rather than assumed through scale, speed, or participation alone. Importantly, the figure also highlights accountability. Human agents remain responsible for how digital outputs are interpreted and mobilised, while digital systems shape, but do not determine, the space of possible reasoning. Argumentation is no longer confined to a single centre or episodic gatekeeping function; instead, it is embedded throughout the network. This signals a mature connectivist configuration in which evaluative reasoning is a continuous practice. Knowledge is produced, challenged, refined, and reintegrated through sustained argumentative engagement across nodes. Argumentation thus becomes the mechanism through which responsibility is distributed without being dissolved.

8.4 Digital Tools for Connectivist Learning

Digital technologies operationalise connectivist principles by enabling rapid connection, collaboration, and feedback. Learning management systems, collaborative modelling environments, simulation platforms, and AI-driven analytics increasingly function as epistemic partners in STEM reasoning. These tools expand cognitive reach by visualising complex systems, aggregating data, and supporting distributed collaboration. Proponents of connectivist learning argue that digital tools fundamentally extend cognition, allowing learners and professionals to reason with representations that would otherwise be inaccessible (Al-Shehri, 2011; Garcia et al., 2015; Siemens, 2005). From this perspective, tool-mediated learning is not auxiliary to cognition but constitutive of it. However, a significant counterargument holds that reliance on digital tools may obscure understanding, displace critical judgment, and introduce unexamined assumptions embedded in algorithms and interfaces (Clarà & Barberà, 2013).

This tension is particularly salient in STEM practice, where algorithmic outputs often carry epistemic authority. Argumentation addresses this risk by repositioning tools as objects of critique rather than sources of unquestioned truth. Practitioners must interrogate data provenance, modelling assumptions, and representational choices, integrating tool outputs into justificatory reasoning rather than deferring to them. In connectivist environments, digital tools therefore function productively only when embedded within argumentative practices that render their limitations visible. Tool fluency without argumentative literacy risks technical efficiency without epistemic responsibility because tools do not reason on our behalf. As we emphasized in Chap. 6, technologies embed values, assumptions, and constraints. Connectivist learning therefore requires critical engagement with tools, interrogating data sources, algorithmic outputs, and visualization choices through

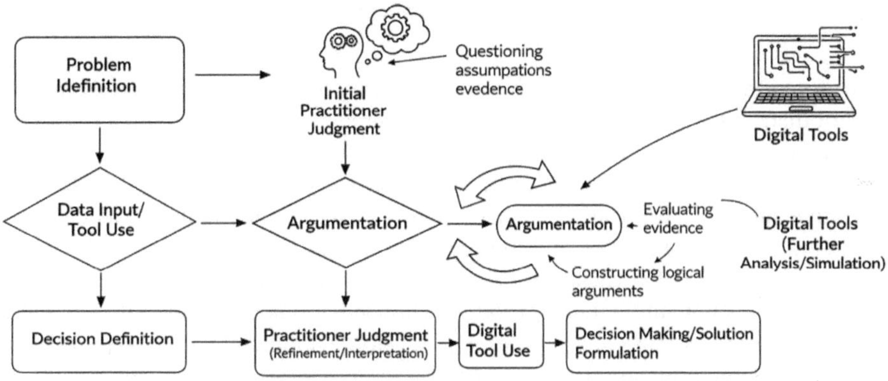

Fig. 8.4 Argumentation-mediated interaction between practitioner judgment and digital tools

argumentation. For practitioners, this means designing workflows where digital tools support, rather than obscure, evidentiary reasoning. For emerging professionals, it means developing fluency not only in tool use, but in tool critique.

Figure 8.4 visually anchors the argument advanced in this section by making explicit how digital tools are integrated into STEM reasoning through argumentation rather than around it. As we move through the workflow from left to right, we see that inquiry begins not with tools, but with problem identification and initial practitioner judgment. This foregrounds the connectivist claim that cognition is distributed, while also affirming that professional reasoning is never tool-neutral or automatic. Before any data are processed, we are already questioning assumptions, framing evidence, and exercising judgment. As the workflow progresses, data input and tool use enter the process, but crucially, they do so within an argumentative space. The central positioning of argumentation illustrates how practitioner judgment and digital tools are brought into dialogue rather than hierarchy. We do not simply accept computational outputs; instead, we cycle between tool-generated analyses and justificatory reasoning. The bidirectional arrows between argumentation and digital tools emphasize this iterative movement, evaluating evidence, constructing logical arguments, and revisiting assumptions embedded in models and simulations.

Seen in this way, the workflow does more than depict a sequence of actions; it clarifies how epistemic authority is negotiated within tool-mediated reasoning. Digital tools enter the inquiry as powerful contributors, yet their outputs acquire meaning only insofar as they are taken up, questioned, and justified within argumentative practice. The recursive movement between argumentation and digital analysis thus makes visible the assumptions, data sources, and representational choices that might otherwise remain implicit. By insisting that tool outputs are continuously routed back through justificatory reasoning, the workflow underscores that cognition may be extended across technologies, but epistemic responsibility is not. What ultimately sustains responsible reasoning in connectivist STEM environments, therefore, is not access to powerful digital tools alone, but the disciplined practice of arguing with and about them.

8.5 Collaborative Knowledge Construction

Collaboration in connectivist environments extends beyond coordinated teamwork toward distributed knowledge construction. Shared documents, open datasets, collaborative simulations, and version-controlled repositories allow participants to co-construct explanations, models, and solutions in ways that exceed individual cognitive capacity. Advocates of connectivist collaboration emphasise its capacity to integrate diverse expertise and to accelerate innovation through collective intelligence (Bell, 2011; Downes, 2012; Al-Shehri, 2011). Yet critics caution that collaboration does not automatically produce knowledge (Garcia et al., 2015); without epistemic norms, collaborative spaces may reinforce dominant voices, suppress dissent, or converge prematurely on consensus (Kuhn, 1993).

In this sense, argumentation provides the structure through which collaboration becomes epistemically productive. Claims must be articulated, evidence made public, counterarguments engaged, and revisions justified. These practices transform collaboration from mere coordination into a disciplined process of knowledge construction aligned with scientific and engineering norms (Osborne et al., 2004). In Chaps. 4 and 5, we emphasized that productive collaboration depends on norms of trust, transparency, and epistemic humility. Importantly, the ethical dimension of collaborative argumentation requires respect for dissent, openness to revision, and recognition of epistemic contributions across hierarchical and disciplinary boundaries. In this sense, professionals must recognize the limits of individual expertise and remain open to revision, a stance that is both cognitive and ethical.

8.6 Creativity and Innovation in Networked Learning

Creativity in STEM is often associated with individual insight or exceptional expertise. Connectivist perspectives reframe creativity as an emergent property of networked interaction, arising through the recombination of diverse ideas, methods, and data sources. Networked learning environments support creativity by enabling practitioners to traverse disciplinary and institutional boundaries, encountering perspectives that challenge established assumptions. Proponents argue that such boundary-crossing is a primary driver of innovation in contemporary STEM fields (Akiri et al., 2020). However, critics note that unstructured exposure to novelty may privilege originality over viability, producing ideas that are inventive but insufficiently grounded in evidence or feasibility (Garcia et al., 2015; Ogunniyi & Iwuanyanwu, 2024). Argumentation plays a generative role in resolving this tension. By subjecting novel ideas to critical scrutiny, counterargument, and evidentiary testing, networks can distinguish productive innovation from speculative divergence. In this way, argumentation does not constrain creativity but channels it toward outcomes that are both imaginative and defensible. Creativity within

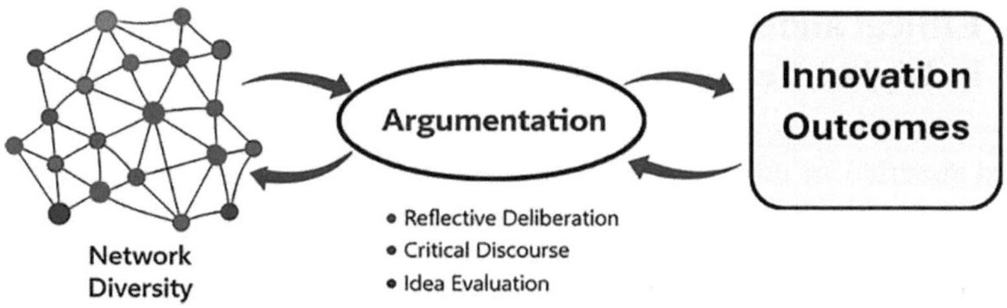

Fig. 8.5 Network diversity, argumentation, and innovation outcomes

connectivist STEM practice is therefore best understood as disciplined emergence, innovation that arises through networked interaction and is stabilised through rigorous argumentative evaluation.

Figure 8.5 makes our earlier conceptual claim visible in one schematic movement. It depicts how networked learning environments support creativity by connecting practitioners across boundaries. As they engage with diverse perspectives, established assumptions are challenged. These encounters are then mediated through reasoned argumentation. On the left, network diversity is represented as a connected web of differently coloured nodes, signalling the variety of perspectives, disciplines, and knowledge sources that practitioners may encounter in networked learning. This visual metaphor aligns with the connectivist claim that creativity is not simply an individual capacity, but an emergent property of interaction across boundaries, a recombination of ideas that arises when diverse voices and resources are connected. The model then directs us toward the centre, where argumentation occupies the pivotal position. The arrows flowing from network diversity into argumentation show that novelty and diversity do not automatically become innovation. Instead, they enter an argumentative space where ideas are tested, challenged, and refined. The three bullet points beneath the central node, reflective deliberation, critical discourse, and idea evaluation, make explicit what the text describes as the "generative role" of argumentation. In this way, the figure confirms that argumentation is not a constraint on creativity but a channel that turns imaginative divergence into defensible outcomes.

Finally, the right-hand section of the model shows innovation outcomes, represented by a lightbulb and growth symbols, emerging from the argumentative process. The arrows indicate that innovation is not the simple result of diversity alone, but of diversity filtered through critical evaluation. This is the visual counterpart to our argument that networked learning can produce creativity that is imaginative yet still grounded in evidence and feasibility. Thus, the figure carries us through the inquiry by showing that creativity in connectivist STEM practice is not the product of novelty alone. Rather, it is disciplined emergence, a process in which diverse network resources generate ideas, and argumentation stabilises and validates them into meaningful innovation.

8.7 Ethical and Responsible Participation in Digital Networks

Ethical practice in connectivist environments cannot be reduced to individual compliance with codes of conduct. Networked STEM systems distribute agency across humans, technologies, and institutions, complicating traditional notions of responsibility. Data-driven decision-making, algorithmic mediation, and platform governance raise ethical questions concerning transparency, bias, accountability, and epistemic justice. Critics of connectivism argue that its emphasis on networks obscures moral responsibility by diffusing agency. This chapter counters that ethical agency in networked environments is exercised through argumentation. Ethical reasoning requires practitioners to articulate values, examine consequences, and justify decisions in ways that remain open to scrutiny and revision (Walton, 2006).

Building on Chap. 7, we argue that ethical agency in STEM requires the ability to recognize how network structures shape participation and outcomes. Argumentation becomes a mechanism for ethical deliberation, enabling stakeholders to surface values, assess risks, and justify courses of action. Importantly, argumentation enables ethical participation by making assumptions visible, exposing power relations within networks, and supporting collective deliberation under uncertainty. In this sense, ethical competence in STEM is inseparable from argumentative competence. Professionals must therefore cultivate ethical literacy alongside technical competence, learning not only how to act, but how to justify action in transparent, accountable ways.

8.8 Argumentation and Decision-Making in Connectivist Environments

Decision-making in connectivist systems is distributed, iterative, and provisional. Professionals draw on networked evidence, consult multiple perspectives, and revise judgments as conditions evolve. Without explicit epistemic structure, such environments risk privileging speed and consensus over rigor and responsibility. Connectivist argumentation may be conceptualised as a recursive cycle:

(i) Problem identification within a networked context
(ii) Mobilisation of distributed evidence
(iii) Critical evaluation of competing claims
(iv) Negotiation of decisions across stakeholders
(v) Ongoing monitoring and revision

This cycle addresses a central critique of connectivism, that it lacks mechanisms for evaluating learning quality, by embedding evaluative judgment directly into networked practice. Argumentation thus resolves, rather than merely responds to,

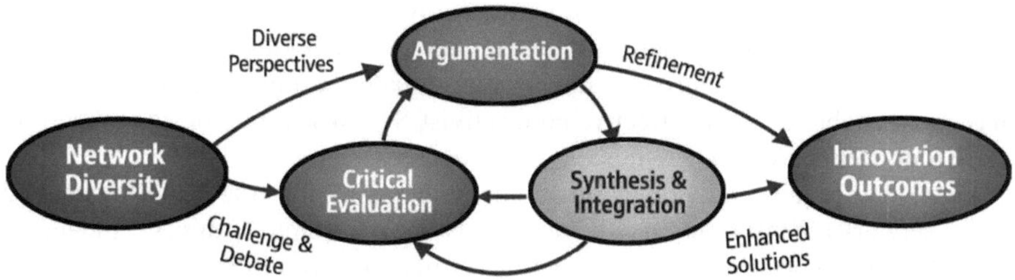

Fig. 8.6 Argumentation as epistemic regulation in connectivist decision-making systems

theoretical objections by supplying the normative criteria connectivism alone does not specify. In Fig. 8.5, we make visible how argumentation operates as a stabilising cycle within connectivist decision-making systems. We begin on the left with network diversity, signalling the distributed and heterogeneous nature of knowledge sources in connectivist environments. Diverse perspectives flow into argumentation, where claims are surfaced, challenged, and made explicit rather than absorbed uncritically.

From here, we trace a recursive movement through critical evaluation and synthesis and integration, illustrating how evidence is weighed, tensions negotiated, and provisional judgments assembled across the network. These stages do not represent a linear pipeline but an iterative loop, one that continually feeds back into argumentation as conditions change and new information emerges. The pathway toward innovation outcomes on the right emphasises that productive decisions arise not from speed or consensus alone, but from disciplined refinement. By foregrounding argumentation as the connective mechanism that regulates movement between diversity, evaluation, and synthesis, the model shows how connectivist decision-making can sustain rigor and responsibility. In this way, the figure visually reinforces our claim that argumentation embeds normative judgment directly into networked practice, addressing longstanding critiques of connectivism's evaluative capacity (Fig. 8.6).

8.9 Designing Connectivist STEM Learning Activities

For educators, curriculum designers, and professional developers, designing connectivist STEM learning activities requires a conceptual shift from organising content sequences to curating epistemic conditions. In connectivist environments, learning does not emerge from transmission but from participation in networks where information, tools, and people interact dynamically (Downes, 2012). Our task, therefore, is not to simplify complexity away, but to design for productive engagement with it. We begin from a philosophical commitment that is often left implicit in STEM education that professional knowledge is provisional, situated, and accountable. From a pragmatist standpoint, knowledge earns its status through use, critique, and revision in response to consequences (Kuhn, 1993). Connectivist design aligns with this view by positioning individuals, including students within authentic problem spaces where answers are not given but must be argued for. This directly challenges transmissive models that equate learning with coverage and mastery of stable content. Effective connectivist STEM learning designs therefore share four defining characteristics.

First, they require individuals, including students to construct, interrogate, and revise networks of information rather than consume pre-curated sources. Students or individuals must be asked not only where information comes from, but why particular sources are trusted, how claims are warranted, and what assumptions remain unexamined. This addresses a central critique of connectivism, namely that access to information can be mistaken for understanding (Clarà & Barberà, 2013). Argumentation functions here as an epistemic scaffold, compelling individuals to justify why certain nodes in their networks carry more weight than others (Toulmin, 1958). Second, connectivist designs embed authentic, ill-structured problems that mirror professional STEM practice (Iwuanyanwu, 2020). Real-world STEM challenges rarely present themselves with clear boundaries, complete data, or single correct answers. By engaging students in open-ended problems, such as socio-technical design dilemmas, data interpretation under uncertainty, or competing model explanations, we invite them into the same epistemic conditions faced by practitioners. This approach resists the temptation to oversimplify complexity for pedagogical convenience and instead treats uncertainty as a productive feature of learning.

Third, digital tools are leveraged not merely for efficiency, but as objects of epistemic inquiry. Simulations, AI-driven analytics, collaborative platforms, and data visualisation tools should be integrated in ways that make their assumptions visible and contestable. We argue that tool use without tool critique produces technically fluent but epistemically vulnerable learners. By contrast, connectivist designs that route tool outputs back through structured argumentation cultivate critical digital judgment, a core professional competence in data-intensive STEM fields (Clarà & Barberà, 2013; Siemens, 2005). Fourth, ethical deliberation must be embedded explicitly rather than treated as an afterthought. Networked STEM practice distributes agency across humans and technologies, complicating responsibility. Designing activities that require learners or individuals to surface values, weigh risks, and justify decisions to multiple stakeholders foregrounds the ethical dimensions of professional judgment (Walton, 2006). From this perspective, ethics is not an external constraint on innovation but an internal condition of responsible practice.

A common counterargument is that such designs risk overwhelming learners, particularly novices, with excessive complexity. We acknowledge this concern. However, the alternative, shielding students from complexity, produces graduates ill-prepared for contemporary STEM realities. The design challenge is therefore not to eliminate complexity, but to sequence epistemic responsibility, gradually increasing students' accountability for network construction, evaluation, and justification. Argumentation once again plays a stabilising role, providing shared norms through which complexity becomes navigable rather than paralyzing. For practitioners preparing students to enter STEM professions, the implication is clear. Connectivist learning activities should be judged not by how smoothly they run, but by how effectively they cultivate reasoned judgment under conditions of uncertainty, diversity, and change.

8.10 Chapter Summary

We have advanced connectivism as a professional epistemic framework suited to the realities of contemporary STEM practice, while also recognising that its coherence and ethical defensibility depend on the regulating force of argumentation. As we participate in digitally mediated and distributed knowledge environments, we encounter a world in which knowledge remains provisional, authority is continually negotiated, and responsibility cannot be surrendered to technologies, platforms, or collective momentum. We have seen that the defining strengths of connectivism, its openness, diversity, adaptability, and distributed nature, also introduce epistemic risks. Without disciplined evaluative practices, networks may reward immediacy over rigor, amplify visibility over credibility, and settle prematurely on consensus without sufficient justification.

We have therefore positioned argumentation as the normative structure that stabilises learning within these environments. Through argumentation, we do not merely connect to knowledge; we examine it, justify it, and remain accountable for the claims we advance and the decisions we make. Participation in networks becomes an epistemic responsibility rather than a passive condition. Creativity emerges not as unbounded novelty, but as disciplined emergence grounded in reasons, evidence, and critique. Collaboration becomes more than coordination; it becomes a shared commitment to interrogating assumptions, negotiating meaning, and refining understanding. Ethical agency, in this sense, resides in our capacity to justify our actions and judgments transparently within the knowledge ecologies we inhabit.

For us as STEM practitioners, this reframes professional competence. We must learn to evaluate networked evidence critically, interrogate the assumptions embedded within digital tools and models, and navigate knowledge claims that cross disciplinary and technological boundaries. For those preparing to enter STEM professions, the task is not simply to form connections, but to justify, evaluate, and revise those connections responsibly. We arrive, then, at a forward-looking recognition: the future of STEM innovation will depend not on technical sophistication alone, but on our capacity to reason carefully, argue responsibly, and act with ethical clarity within complex and evolving knowledge networks. When disciplined through argumentation, connectivism becomes more than a description of how we learn. It becomes a framework for who we must become. We are called to act as responsible agents who shape knowledge, professional practice, and technological futures. This responsibility requires intellectual integrity and a sustained commitment to moral purpose.

8.10.1 From Networked Knowing to Responsible Becoming

In Chap. 8, we brought our inquiry into STEM professionalism to its most expansive horizon by examining connectivism as an epistemic framework for learning and acting within distributed, digitally mediated knowledge environments. We showed that participation in such networks places new demands on practitioners and students alike: not only to access and share information, but to evaluate its credibility, justify its use, and remain accountable for the decisions it informs. In this context, we argued that argumentation becomes indispensable. It is argumentation that disciplines networked learning, transforming connection into understanding, collaboration into critique, and participation into responsibility. Through argumentation, we come to see that knowledge is not merely encountered, but actively constructed, examined, and defended within communities of inquiry and practice.

This insight brings us to a deeper realisation that extends beyond any single framework or chapter. If connectivism reveals the conditions under which contemporary STEM knowledge is formed, argumentation reveals the ethical and intellectual responsibilities that accompany our participation in those conditions. We are no longer merely learners navigating networks or professionals applying established methods; we become agents whose judgments shape the integrity of knowledge itself and, by extension, the socio-technical futures that knowledge makes possible. Our capacity to argue well becomes inseparable from our capacity to act wisely.

It is at this philosophical threshold that the concluding chapter takes up its final task. Having examined argumentation as epistemic practice, scientific reasoning, engineering design, and networked learning, we now confront its enduring significance for uncertain and ethically charged STEM futures. The conclusion does not seek to summarise what has been learned as much as to clarify what must be sustained. It invites us to recognise argumentation as more than a pedagogical strategy or professional skill; it is a way of inhabiting STEM with intellectual humility, ethical seriousness, and a commitment to accountable reasoning. As we bring this handbook to a close, we pause to reflect on a deeper question. It is no longer only about how STEM professionals learn, design, and connect. We ask who they must become. They are called to reason thoughtfully with others. They must justify their actions with care and integrity. They are tasked with shaping technological futures that are not only innovative, but also humane, just, and sustainable. In this responsibility, learning, argumentation, and ethical practice converge. The question of becoming, we see, is inseparable from the question of doing well in the world.

8.11 Reflection Questions for Practice

Rather than reflective prompts, the following provocations are intended as professional self-audit questions that practitioners and emerging professionals can revisit as their roles, networks, and responsibilities evolve:

1. How do the structures of your professional networks enable or constrain the quality of evidence you encounter and prioritise?
2. Where do digital tools in your practice support epistemic rigor, and where might they obscure assumptions, values, or uncertainties?
3. How are ethical considerations surfaced, debated, and justified within your professional networks, particularly when decisions must be made under time or data pressure?
4. In what ways does argumentation function explicitly in your professional decision-making, and where does it remain implicit or underdeveloped?
5. How prepared are you to revise judgments as new evidence, perspectives, or stakeholders enter the network?

References

Akiri, E., Tal, M., Peretz, R., Dori, D., & Dori, Y. J. (2020). STEM graduate students' systems thinking, modeling and scientific understanding—The case of food production. *Applied Sciences, 10*(21), 7417. https://doi.org/10.3390/app10217417

Al-Shehri, S. (2011). Connectivism: A new pathway for theorising and promoting mobile language learning. *International Journal of Innovation and Leadership in the Teaching of Humanities, 1*(2), 10–31.

Bell, F. (2011). Connectivism: Its place in theory-informed research and innovation in technology-enabled learning. *The International Review of Research in Open and Distributed Learning, 12*(3), 98–118.

Clarà, M., & Barberà, E. (2013). Learning online: Massive open online courses (MOOCs), connectivism, and cultural psychology. *Distance Education, 34*(1), 129–136. https://doi.org/10.1080/01587919.2013.770428

Downes, S. (2012). *Connectivism and connective knowledge. Essays on meaning and learning networks*. National Research Council Canada.

Garcia, E., Elbeltagi, I., Brown, M., & Dungay, K. (2015). The implications of a connectivist learning blog model and the changing role of teaching and learning. *British Journal of Educational Technology, 46*(4), 877–894. https://doi.org/10.1111/bjet.12184

Goldie, J. G. S. (2016). Connectivism: A knowledge learning theory for the digital age? *Medical Teacher, 38*(10), 1064–1069. https://doi.org/10.3109/0142159X.2016.1173661

Iwuanyanwu, P. N. (2020). Nature of problem-solving skills for 21st century STEM learners: What teachers need to know. *Journal of STEM Teacher Education, 55*(1), 27–40. https://doi.org/10.30707/JSTE55.1/MMDZ8325

Kuhn, D. (1993). Science as argument: Implications for teaching and learning scientific thinking. *Science Education, 77*(3), 319–337. https://doi.org/10.1002/sce.3730770306

Ogunniyi, M., & Iwuanyanwu, P. N. (2024). Analysis of teachers' perspectives towards the use of IKS to improve STEM education for sustainable development. *African Journal of Research in Mathematics, Science and Technology Education, 28*(3), 319–329. https://doi.org/10.1080/18117295.2024.2352980

Osborne, J., Erduran, S., & Simon, S. (2004). Enhancing the quality of argumentation in school science. *Journal of Research in Science Teaching, 41*(10), 994–1020. https://doi.org/10.1002/tea.20035

Shrivastava, A. (2018). Using connectivism theory and technology for knowledge creation in cross-cultural communication. *Research in Learning Technology, 26*, 2061. https://doi.org/10.25304/rlt.v26.2061

Siemens, G. (2005). Connectivism: A learning theory for the digital age. *International Journal of Instructional Technology and Distance Learning, 2*(1), 3–10.
Toulmin, S. (1958). *The uses of argument*. Cambridge University Press.
Walton, D. (2006). *Fundamentals of critical argumentation*. Cambridge University Press.

Chapter 9
Recapitulation and Conclusion

In this concluding chapter, we move deliberately from framework to praxis. We ask what it means for STEM practitioners and emerging professionals to live argumentation as a form of critical practice, one that shapes decisions, responsibilities, and futures in ethically charged STEM landscapes. Across Chaps. 1, 2, 3, 4, 5, 6, 7, 8, we have advanced a coherent and cumulative argument that argumentation is the epistemic architecture that sustains STEM reasoning, learning, and professional judgment. From the opening chapters that positioned argumentation as a cognitive, social, and epistemic practice, to later chapters that examined argumentation enactments, assessment, professional learning, and systemic constraints, we have consistently argued that STEM competence without argumentative capacity is incomplete. As Toulmin's model of argument, Kuhn's work on epistemic development, and contemporary science education scholarship remind us, claims gain meaning only through warrants, evidence, and critical scrutiny.

In Chaps. 1 and 2, we established argumentation as the connective tissue of STEM practice, linking evidence, theory, modelling, ethics, and decision-making. Chapter 3 clarified what arguments are, and what they are not, laying the conceptual groundwork for principled reasoning. Chapter 4 deepened this foundation by articulating argumentation as an epistemic practice through which claims are constructed, warranted, challenged, and refined. Chapters 5, 6, 7 demonstrated how these practices unfold in technologically mediated environments, scientific inquiry, and engineering design. Chapter 8 extended the framework into connectivist and networked contexts, arguing that argumentation regulates knowledge flow, coherence, and responsibility. This conclusion tightens these threads and turns explicitly toward praxis. We ask how the frameworks developed across the book become lived forms of professional reasoning for STEM practitioners and students preparing to enter STEM professions. Our claim here is not additive but integrative. The value of argumentation lies precisely in its capacity to move fluidly between epistemic rigor, ethical deliberation, and practical action. This integrative perspective sets the stage

P. N. Iwuanyanwu, *Empowering STEM Thinkers Through Argumentation*, Contemporary Trends and Issues in Science Education 69,
https://doi.org/10.1007/978-3-032-27906-4_9

for visualizing argumentation in action, tracing how the conceptual frameworks developed in earlier chapters manifest across nested scales of STEM practice, from learning environments to professional contexts.

In Fig. 9.1, we are carried across the arc of the book, moving deliberately from epistemic framework to lived STEM practice. We see argumentation not as a single skill or isolated classroom activity, but as a practice that travels, operating across three nested and mutually informing scales of STEM work. We begin in STEM learning environments, where argumentation is first encountered, rehearsed, and disciplined. This is the familiar terrain developed in the early chapters where we learned how to advance claims, how to coordinate evidence and warrants, and how to engage in critique. At this level, argumentation functions as an epistemic apprenticeship, rendering reasoning visible, contestable, and improvable. From this grounding, the inquiry extends into professional practice and design, where argumentative resources are recontextualised under conditions of uncertainty, constraint, and consequence. The same structures of justification and critique that supported learning now organise engineering design, scientific judgment, and ethical deliberation. Practice, here, is not a departure from learning but its continuation. In other

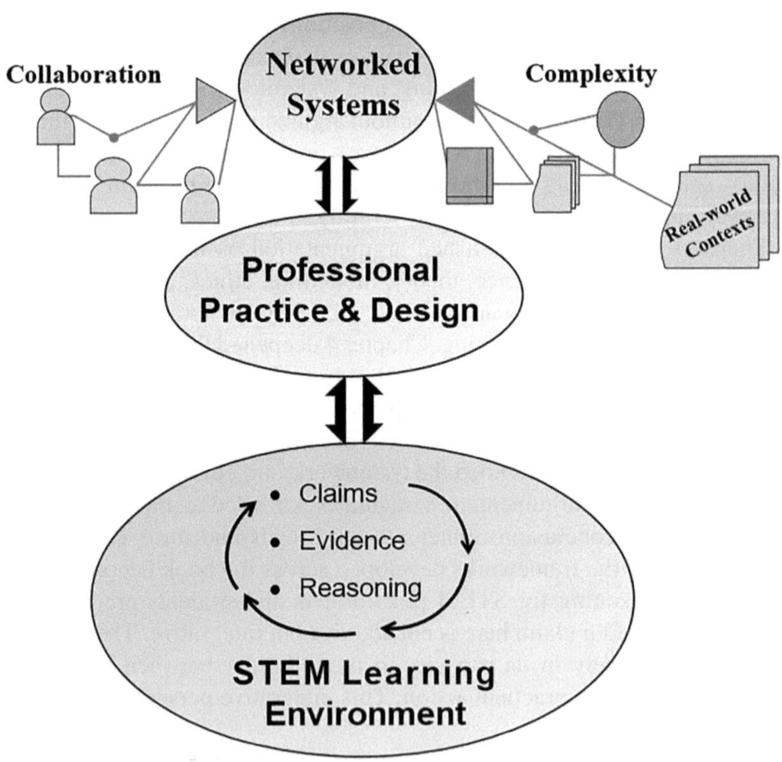

Fig. 9.1 Argumentation across STEM learning, professional practice, and networked systems

words, what is cultivated in educational settings becomes the means through which professionals reason, decide, and take responsibility in real-world contexts.

The inquiry widens further into networked and connectivist systems, where argumentation no longer resides solely within individuals or professions, but circulates across policies, institutions, digital infrastructures, and socio-technical networks. As argued in Chap. 7, argumentation at this scale regulates the flow, coherence, and legitimacy of knowledge itself, shaping how claims are taken up, challenged, and revised across distributed communities. What counts as sound reasoning is thus negotiated not only locally, but systemically. Movement across these levels is neither linear nor hierarchical. Instead, it is iterative and recursive, looping continuously between learning, practice, and systems.

Educational contexts shape professional reasoning; professional practices expose new epistemic and ethical demands; and systemic conditions, in turn, reframe the possibilities for learning and action. Tracing these movements allows us to see that argumentation is not something we leave behind as we move "up" into professional life. It is the connective tissue that keeps epistemic rigor, ethical judgment, and practical action in sustained dialogue across scales. In this way, the framework developed across the book turns decisively toward praxis, illustrating how argumentation becomes a lived practice—learned, enacted, and sustained, as STEM practitioners navigate increasingly complex and ethically charged landscapes.

9.1 Argumentation as Epistemic Apprenticeship and Critical Praxis in STEM Professions

Critical praxis requires more than technical mastery. It demands reflective judgement, ethical reasoning, and the capacity to engage with uncertainty. Drawing on Freire's conception of praxis as reflection and action upon the world, we argue that argumentation operationalises critical praxis in STEM. Through argumentation, practitioners interrogate assumptions, surface values embedded in models and technologies, and negotiate competing explanations or solutions. For practicing engineers, scientists, technologists, and educators, this means recognising that professional decisions, whether designing infrastructure, interpreting data, or implementing curricula, are always value-laden. Habermas's concept of communicative rationality reinforces this perspective. It shows that rationality in professional STEM contexts emerges not through instrumental reasoning alone, but through dialogic justification that is open to scrutiny and deliberation. A common counterargument is that time pressures and performance metrics in STEM fields leave little room for extended argumentation. We acknowledge this tension. However, the literature on expert practice shows that tacit argumentation already underpins high-level decision-making. Making it explicit strengthens rather than slows professional judgement. Argumentation, then, is not an add-on. It is a discipline for thinking well under constraint.

In Chap. 4, we positioned argumentation not as debate for its own sake, but as an apprenticeship into disciplinary ways of knowing. We argued that when individuals, including students engage in constructing and critiquing arguments, they participate in the same epistemic practices that define scientific inquiry, mathematical justification, and technological decision-making. This conclusion reaffirms that stance and extends it forward. For STEM educators, argumentation-rich classrooms function as microcosms of professional practice. The careful coordination of claims, evidence, warrants, and rebuttals mirrors how knowledge is stabilised within disciplinary communities.

Importantly, Chap. 4 emphasised that uncertainty is not a deficit to be eliminated but a condition to be reasoned with. In this sense, argumentation cultivates epistemic resilience: students learn to remain intellectually engaged when answers are provisional, data are incomplete, or models conflict. A persistent counterargument suggests that such epistemic openness risks confusing learners or slowing curricular progress. Our response, grounded in both philosophical and empirical traditions discussed earlier in the book, is that intellectual discomfort is not a pedagogical failure but a catalyst for deeper understanding. Argumentation transforms uncertainty into deliberation and indecision into disciplined judgment. It is precisely this cultivation of epistemic resilience in the classroom that Fig. 9.2 captures, showing how argumentation functions as an apprenticeship into disciplinary reasoning.

The figure guides us step by step, tracing how everyday claims are gradually disciplined through the coordination of evidence, warrants, and critique, ultimately shaping the forms of judgment recognised and stabilised within STEM communities. We begin by observing the informal ways in which claims are often made in everyday contexts and then follow how these claims become increasingly structured as learners engage in iterative reasoning and critique. As reasoning develops, claims must be justified and rendered open to evaluation, reflecting the norms that govern disciplinary communities. Critique, rather than signalling failure, marks a

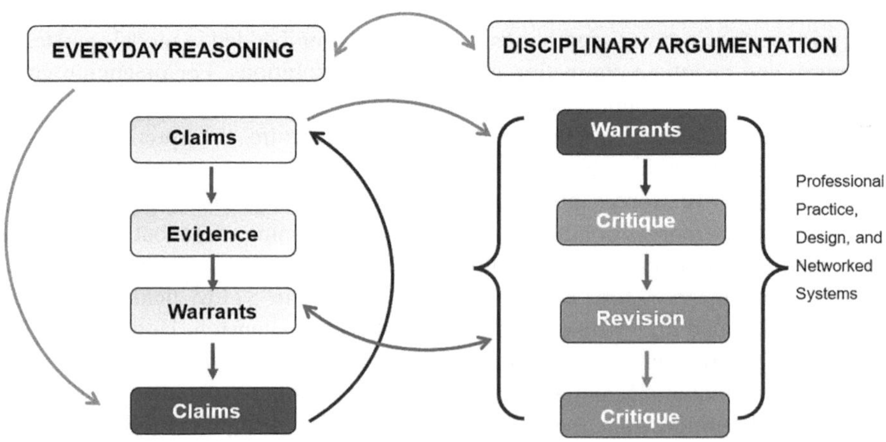

Fig. 9.2 Argumentation as epistemic apprenticeship

deepening of epistemic engagement, reinforcing the idea that uncertainty and provisional understanding are essential components of authentic reasoning practice. Revision, rather than closure, emerges as a central moment in this journey. We are shown how argumentation is inherently iterative, looping back on itself as claims are refined in response to challenge. This recursive movement aligns with the conception of praxis developed earlier in the chapter: reflection and action remain inseparable, and judgment is continually reshaped by encounter with uncertainty.

Importantly, the trajectory does not terminate within educational settings. The same argumentative scaffolds that support learning echo outward into professional practice and design, where decisions must be justified under conditions of constraint, risk, and ethical consequence. What is rehearsed in classrooms thus anticipates the epistemic demands of STEM professions, preparing individuals to reason responsibly when answers are provisional and stakes are high. By guiding us through this progression, the inquiry reinforces a central claim of the book: argumentation is not an add-on to STEM learning or professional work, but the means through which epistemic rigor, ethical reflection, and practical action are woven together. In this way, argumentation functions simultaneously as apprenticeship and praxis, how individuals learn to think, decide, and act well within the complex realities of STEM practice.

9.2 Professional Learning, Design Thinking, and Ethical Deliberation

For STEM educators, this handbook has positioned argumentation as both a pedagogical approach and a professional stance. Teaching for argumentation requires intentional task design, assessment practices that value reasoning over recall, and classroom cultures that legitimise uncertainty and disagreement. Drawing on socio-constructivist learning theory and empirical studies reviewed in earlier chapters, we reiterate that students learn STEM best when they are treated as sense-makers rather than answer-producers. We also recognise the structural challenges educators face, which include high-stakes assessments, overloaded curricula, and limited professional development. Yet, as we have argued, even small shifts, such as asking students to justify claims, critique models, or compare explanations, can reorient classrooms toward argumentative practice. For students preparing to enter STEM professions, argumentation serves as an apprenticeship into disciplinary ways of knowing. Learning to argue well is learning to participate legitimately in STEM communities of practice. In Chap. 7, we demonstrated that engineering design thinking is inherently argumentative, ethical, and socially situated. Decisions about optimisation, safety, sustainability, and feasibility are never purely technical; they are negotiated through justification, trade-offs, and value-laden reasoning.

For practicing professionals, argumentation functions as a mode of reflective accountability. Design documentation, risk disclosure, and stakeholder engagement, examined in Chap. 7, are not administrative afterthoughts but argumentative acts through which practitioners make their reasoning visible and contestable. In this sense, professional expertise is marked not only by what one decides, but by one's capacity to justify decisions in light of evidence, constraints, and ethical consequences. We also acknowledge a pragmatic counterpoint often raised in industry contexts: that explicit ethical argumentation may appear to conflict with efficiency or innovation speed.

Yet, as we argued in Chap. 7, unexamined optimisation frequently externalises risk and defers ethical costs to society. Argumentation does not constrain creativity; it disciplines it, ensuring that innovation remains accountable to human and ecological futures. If argumentation is what disciplines creativity and anchors professional judgment to ethical and evidentiary accountability, we must then ask how this practice is enacted within the everyday rhythms of design work itself. How, in concrete terms, do epistemic, creative, and ethical arguments surface as engineers move from defining problems to proposing, testing, and communicating solutions? It is this question, how argumentation becomes operative within the design process, that guides our attention to the engineering design cycle.

In Fig. 9.3, we make visible how argumentation organises professional learning across the engineering design cycle. Rather than treating design as a linear sequence of technical steps, we are invited to trace how argumentation quietly but persistently organises movement across the cycle. As we move through the design process, we begin to ask: where do claims emerge, what counts as evidence at each moment, and how are reasons weighed against constraints, values, and consequences? At the outset, problem framing draws us into epistemic argumentation. Here, the central

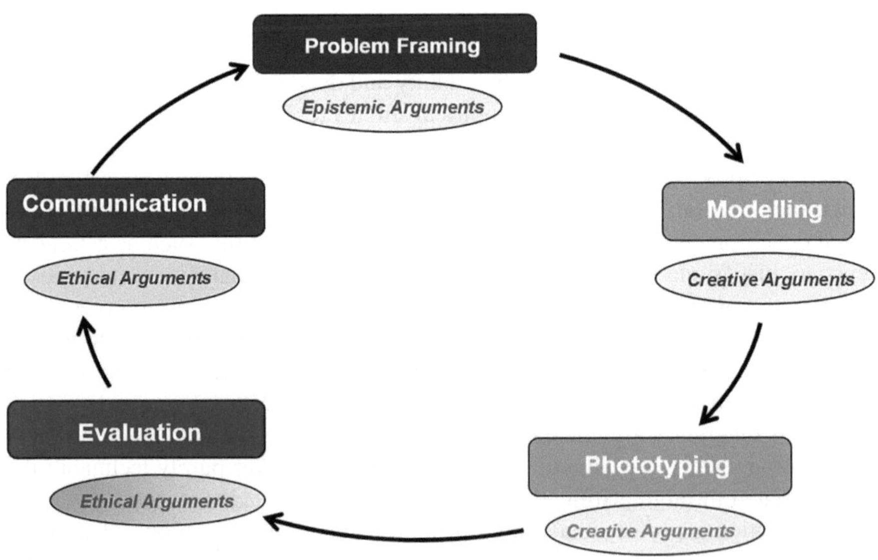

Fig. 9.3 Argumentation across the engineering design cycle

question is not simply what problem should be solved, but whose problem it is, how it is defined, and what assumptions shape that definition?

We find ourselves justifying the boundaries of the problem space, negotiating priorities, and deciding which forms of knowledge are treated as legitimate. These early argumentative moves already carry ethical weight, as exclusions and emphases subtly anticipate downstream consequences. As attention shifts toward modelling, creative argumentation comes to the fore. Competing representations, assumptions, and simplifications must be defended in relation to purpose rather than correctness alone. We are prompted to consider why one model is preferable to another, how uncertainty is managed, and what is gained or lost through abstraction. In this phase, argumentation becomes a resource for imaginative yet disciplined exploration, allowing creativity to be articulated, tested, and refined.

Prototyping intensifies this dialogue between ideas and material constraints. Emerging artefacts begin to "answer back," requiring designers to justify modifications, compromises, and design choices in light of performance, feasibility, and safety. Here, epistemic and creative arguments intersect, while ethical considerations begin to surface more visibly as risks, users, and contexts become harder to ignore. During evaluation, ethical argumentation moves to the foreground without displacing technical reasoning. Decisions about success, optimisation, and failure demand explicit justification against criteria that extend beyond efficiency or cost. We are compelled to ask whether what works ought to be implemented, for whom, and under what conditions. Evidence is weighed not only for accuracy but for consequence.

Finally, communication consolidates the cycle as an explicitly argumentative practice. Design decisions are rendered visible and contestable through reports, presentations, and stakeholder engagement. Justification here is not an endpoint but an invitation to scrutiny, revision, and accountability. By making reasoning public, professional learning is sustained across communities of practice. Taken together, this progression guides us through the design landscape much like a navigational system: not by repeatedly announcing where we started, but by orienting our judgment at each turn. What becomes clear is that engineering design is held together not by procedures alone, but by ongoing argumentation that integrates knowledge, creativity, and ethics. In this sense, professional competence is less about arriving at a solution and more about sustaining a defensible path through complex, value-laden terrain.

9.3 Argumentation, Ethics, and the Moral Work of STEM Futures

Ethical STEM futures depend on how present-day practitioners reason, justify, and contest decisions. Whether addressing climate change, artificial intelligence, public health, or energy transitions, STEM professionals operate in morally complex terrains. Argumentation provides a structure for ethical deliberation by requiring practitioners to articulate assumptions, evaluate consequences, and respond to counterarguments. Philosophical traditions in ethics, ranging from consequentialism to virtue ethics, converge on the importance of reasoned justification. In STEM contexts, argumentation becomes the bridge between technical evidence and ethical responsibility. We have shown throughout the handbook that students who engage in structured argumentation develop not only conceptual understanding but also epistemic humility and moral awareness. Critics may argue that ethics should be taught separately from core STEM content. But separating ethics from practice risks producing professionals who can optimise systems without questioning their purposes. Argumentation embeds ethical reasoning within everyday STEM problem-solving, making ethical reflection habitual rather than episodic.

Across the book, we have insisted that ethics is not an external add-on to STEM practice but an internal dimension of reasoning itself. Chapter 5 highlighted this point in relation to AI, automation, and data-driven systems, where computational outputs can mask normative assumptions. Chapter 6 reinforced it through scientific inquiry, where methodological choices shape how knowledge enters the world. In this concluding synthesis, we frame argumentation as the moral work of STEM practice. Through argumentation, practitioners articulate not only what can be done, but what ought to be done. Ethical deliberation becomes visible through the weighing of consequences, the interrogation of assumptions, and the consideration of alternative futures. Importantly, argumentation sustains moral agency under conditions of uncertainty, conditions that increasingly define contemporary STEM work.

9.4 Professional Identity, Agency, and the Argumentative Self

A distinctive motif across this handbook has been the cultivation of the STEM thinker as a reflective, ethical agent rather than a mere technical problem-solver. Argumentation is central to this professional identity. To argue well is to accept responsibility for one's claims, to remain open to critique, and to recognise that knowledge is always provisional and revisable. For students preparing to enter STEM professions, learning to argue is learning to belong. It marks a transition from reproducing authorised knowledge to participating in knowledge-building enterprises. For experienced practitioners, argumentation supports professional agency in contexts shaped by institutional pressure, commercial interests, and socio-political constraints. It equips professionals to resist uncritical compliance and to justify dissent when evidence or ethics demand it. In this regard, argumentation is inseparable from professional identity.

To see oneself as a STEM professional is to see oneself as someone who can defend claims, evaluate evidence, and respond responsibly to critique. Throughout this book, we have emphasised that argumentative competence supports professional agency, the capacity to act with justification rather than compliance. This is particularly important in contexts where STEM professionals operate under institutional, political, or commercial pressures. Argumentation equips practitioners to resist unethical directives, challenge flawed assumptions, and advocate for socially just outcomes. Argumentation for ethical STEM futures must therefore be inclusive, attentive to power, and open to diverse ways of knowing.

9.5 From Individual Practice to Systemic Responsibility

While much of this book has focused on classrooms and professional practice, its implications extend to systems of STEM education and governance. Accreditation standards, policy decisions, and professional codes rely implicitly on argumentative reasoning. Embedding argumentation explicitly within these structures strengthens transparency, accountability, and trust. We caution, however, against reducing argumentation to a procedural checklist. As we have argued throughout, argumentation is a living practice, shaped by disciplinary norms, cultural contexts, and evolving technological conditions. Its power lies in its capacity to hold together epistemic rigor, ethical responsibility, and creative possibility. Recognizing argumentation as a living, context-sensitive practice at the classroom and professional level invites us to explore how it extends into the broader systems that govern STEM education, shaping policies, standards, and professional norms.

Building on this perspective, Fig. 9.4 invites us to explore how argumentation functions across nested scales, connecting learning environments, professional contexts, and the broader systems that sustain STEM education. With this orientation, we begin to trace how argumentation operates not as an isolated skill or local practice, but as a connective logic that coordinates action, responsibility, and accountability across scales. As we move through this landscape, we are prompted to ask how argumentation in STEM learning environment travels outward. When students justify claims, critique evidence, and revise explanations, these practices do not end at the classroom door. They echo into professional norms, informing how future practitioners document decisions, respond to critique, and engage with uncertainty. We start to see how habits of reasoning cultivated in learning environments become resources for professional judgment rather than disposable academic exercises.

Following this trajectory further, our attention is drawn to professional practice as a mediating space. Here, justification is formalised through design reports, standards compliance, and ethical review. We ask how arguments are stabilised into conventions, and how critique is managed within institutional constraints. At this level, argumentation functions as a mechanism for reflective accountability, allowing expertise to be questioned without collapsing into arbitrariness. Yet we also begin to wonder: what happens when these professional arguments encounter policy frameworks and governance structures? As the inquiry broadens, policy and accreditation emerge not as distant regulators, but as participants in ongoing argumentative cycles. Standards, codes, and regulations rely on claims about quality, safety,

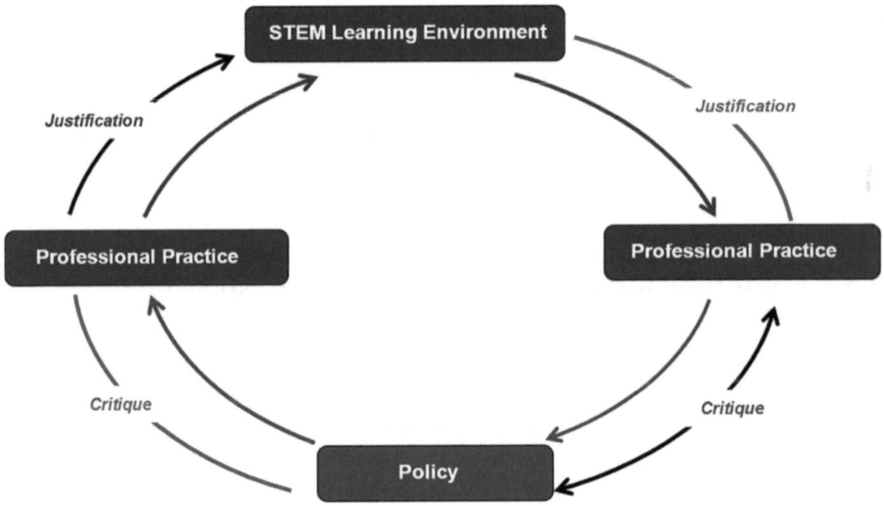

Fig. 9.4 Argumentation as a systemic regulator

and public good, all of which must be defended, revised, and justified over time. We are invited to consider how feedback from practice challenges policy assumptions, and how policy decisions, in turn, reshape what counts as legitimate argumentation in classrooms and professions alike.

Moving through these interconnected spaces, we notice that no single level operates in isolation. Justification flows upward, critique circulates laterally, and revision loops back downward. The system remains dynamic precisely because argumentation is not reduced to compliance. Instead, it functions as a living regulator, holding epistemic rigor, ethical responsibility, and creative possibility in productive tension. Seen this way, the journey we are guided through reframes responsibility in STEM education. Accountability is no longer located solely in individual actors, but distributed across systems that must continually explain, defend, and revise their decisions. What emerges is an understanding of argumentation as the connective tissue of systemic trust, quietly coordinating classrooms, professions, and policies without needing constant reminders of where the journey began.

9.6 Argumentation for Uncertain and Ethical STEM Futures

We have journeyed from the foundations of argumentation, through the practices of rigorous scientific thinking, into the ethical and creative demands of engineering design, and finally into the networked, digitally mediated landscapes of connectivist learning. Along the way, we have seen that knowledge is never static; it is made, tested, questioned, and defended through our engagement with others. Argumentation makes reasoning visible and accountable. Scientific practices give form to inquiry and evidence. Design thinking asks us to balance possibility with responsibility. Connectivism reminds us that our learning and action unfold within complex networks, where every connection carries opportunity and responsibility alike.

Through these chapters, we have come to understand that STEM professionalism is inseparable from ethical and epistemic judgment. Knowledge does not exist in isolation; it emerges in dialogue, in critique, and in the careful negotiation of competing claims and values. Creativity is disciplined by evidence, collaboration is sustained by critique, and ethical agency resides in our capacity to justify decisions transparently within the systems we inhabit. Each chapter has built on the previous, moving from understanding argumentation as a cognitive and cultural practice, to seeing scientific reasoning and design as tools for shaping the world responsibly, to recognising the profound responsibilities embedded in networked learning environments.

In the face of uncertainty, complexity, and ethical urgency, the future of STEM will not be secured by skill alone. No fixed curriculum, set of rules, or body of knowledge can fully prepare us for this future. What endures is our ability to reason with others, to weigh evidence carefully, to justify our decisions, and to remain open to revision. We invite educators to create learning environments where argumentation becomes a habit of thought, a professional ethic, and a source of empowerment. We ask practitioners to embrace argumentation as a guiding ethic, shaping judgment even under pressure, and act as responsible agents shaping technological futures that are not only innovative, but humane, just, and sustainable. We call on students and emerging professionals to see argumentation not as an obstacle, but as a source of intellectual empowerment and moral clarity. In nurturing these capacities, the practice of STEM and the work of becoming are inseparable, and the capacity to reason well becomes both a compass and a moral imperative, guiding us through the uncertainties and possibilities of the worlds we help to create.

MIX
Papier aus verantwortungsvollen Quellen
Paper from responsible sources
FSC® C105338

If you have any concerns about our products,
you can contact us on
ProductSafety@springernature.com

In case Publisher is established outside the EU,
the EU authorized representative is:
Springer Nature Customer Service Center GmbH
Europaplatz 3, 69115 Heidelberg, Germany

Printed by Libri Plureos GmbH
in Hamburg, Germany